Paradise Regained

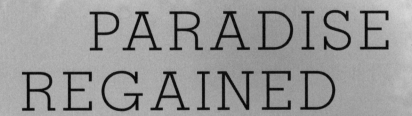

PARADISE
REGAINED

The Regreening of Earth

· Les Johnson · Gregory L. Matloff · C Bangs ·

Published in the United States by Copernicus Books,
an imprint of Springer Science+Business Media.

Copernicus Books
Springer Science+Business Media
233 Spring Street
New York, NY 10013
www.springer.com

Library of Congress Control Number:
2009940203

Manufactured in the United States of America.
Printed on acid-free paper.

ISBN 978-0-387-79985-8   e-ISBN 978-0-387-79986-5

Contents

Foreword   vii
Acknowledgments   xi
Synopsis and Chapter Summaries   xiii
Author Biographies   xix
Introduction: Welcome to the Present   xxiii

1. Space Utilization: A Moral Imperative   1

Part I. Mythical Paradise   9
2. Fire: Formation of Earth and the Solar System   11
3. Earth Before Man: Utopia or Nightmare?   21

Part II. Paradise Lost?   27
4. The Environmental Dilemma: Progress or Collapse?   29
5. Exploding Population   39
6. Climate Change   49
7. Vanishing Life   59
8. Diminishing Energy   67
9. Humans Before the Industrial Age: A Desirable Ecological Goal?   79

Part III. Paradise Regained   87
10. Raw Materials from Space   89
11. Power from the Sun   107
12. Environmental Monitoring from Space   119
13. Protecting Earth   133
14. Mitigating Global Warming   145
15. Settling the Solar System   157
16. Paradise Regained: An Optimistic Future   169

Afterword: Why Space Advocates and Environmentalists Should Work Together   175

Index   177

Dedicated to

My children, Carl and Leslie.
The future I am so dedicated to preserving belongs to them.
Les Johnson

Mother Nature, may she always be whole!
Gregory L. Matloff

The Earth and my parents, who bequeathed to me
their love for her.
C Bangs

Foreword

Gloom and doom sells. That salient point is made by the authors and, for me, is reinforced every time I sit down in front of my favorite cable news channel. Mass shooting at a high school in Iowa. The glaciers are still melting. Terrorists attack a hotel in Pakistan. The whales are going extinct. We hear about loose nukes, lunatics in power, corrupt politicians, child abuse, famines in Africa, tidal waves in Thailand. A woman in the Middle East is raped; then she is accused of immorality for allowing the attack to happen, and is murdered. Racial conflict shows up in Cambridge, MA. News arrives every day that the United States may sink under its accumulation of debt. And we all know the world is running out of oil.

Despite all this, the authors point that things have been worse. At the turn of the last century, people in the U.S. were, on average, not living much past forty. One thinks of the pre-Civil Rights era, of the Depression, of the nuclear confrontation between the superpowers over Cuba. I can recall, during those years, watching workers build an overpass on the Baltimore-Washington Expressway, and wondering why they bothered. At that time, an eventual all-out war seemed inevitable.

Paradise Regained has three authors. Gregory L. Matloff, a physicist at the New York City College of Technology; his wife, C Bangs, a Brooklyn artist; and NASA physicist Les Johnson.

Matloff is convinced that science and intelligence will eventually win out over the lunacy that has plagued mankind since Herodotus was writing history. He has published more than a hundred scientific papers, and is the author of, or collaborator on, seven other science books.

Bangs's work might be said to be inspired by starlight. She uses Gaia, goddesses, and the night sky to portray various aspects of the cosmos. Her conversations with her husband about life, death, and the universe provide much of the insight on display in her art.

Les Johnson has managed various in-space propulsion technology projects for NASA. He's worked on a tether experiment using Earth's magnetic field for propulsion. And he has twice received NASA's Exceptional Achievement Medal.

In Paradise Regained, the authors recognize the severity of the problems that humanity faces, and they don't pretend that technological break-throughs alone will be enough to save us. But it is clear that, without the technology, we are headed for a catastrophe.

And there is reason for optimism. After all, life is getting better. We live more comfortably than our grandparents did. It is now possible that African-American neighbors can show up in the neighborhood without unduly alarming those who really liked the nineteenth century. We now have the internet; we are showing signs of getting rid of tobacco, and medical science has come a long way. And we've developed a global sensitivity that we never really had before.

The authors suggest that the growing awareness that we share the same world, and that it has its limitations, began with those first photos of Earth taken from the moon, the pictures of the fragile blue world drifting through an endless sky. They're probably right. If the human race ever does really coalesce into a family, I suspect those pictures will be hanging near the front door of the family estate.

Will it ever happen? Bangs, Matloff, and Johnson think it will. And, in Paradise Regained, they lay out a plan to make it possible. If we choose to make the effort, to collaborate, to work together, here, they say, are the tools we will need. Here's how to deal with the inevitable energy shortages that are on the horizon, and do it in a way that does not wreck the ecosystem. They point out that we have a virtually infinite supply of clean energy available, compliments of the sun. All we need do is make the investment to harness it.

Here also is a technique for getting rid of the pollution caused by various manufacturing industries. And we might also want to take a long look at the dangers presented by near-Earth asteroids. A single rock, a mile or so in diameter, could put the lights out for all of us. Permanently. Most people shrug at the scenario. They would ask how many of them are there? The answer, unfortunately, is that they are numerous. And they are all over the place. Two weeks ago, as I write this, something very large crashed into Jupiter. We never saw it coming.

Then there are the issues of global warming. And conservation of resources. With world population at six billion and climbing, recycling aluminum cans, planting more trees, and turning the air conditioner down a notch won't get the job done. Won't come close. We're faced with serious problems, and eventually both will reach crisis stage.

Do the authors have a plan? You bet. It's not a solution we could manage today, because we don't have the technology yet. But there's time. If we act. If we avoid our usual propensity of waiting until the flood waters are running into the valley. And that's the problem with the plan. It will require political will and advance planning. And maybe, most important of all, imagination.

I'd like very much to see a copy of Paradise Regained placed in the hands of leaders, and talk show hosts, around the world. The rest of us will be able to profit by it too.

Jack McDevitt
Nebula-winning author of *Time Travelers Never Die*

Acknowledgments

We would like to thank Ken Roy, Robert Kennedy, and David Fields for their contributing Chapter 14, Mitigating Global Warming. We are blessed to have such technically innovative thinkers as both colleagues and friends.

Thanks also to Mitzi Adams and Sam Lightfoot for providing their technical expertise in reviewing Chapters 4 and 6.

We appreciate our colleagues, friends, and students, who have been a constant source of inspiration.

I (Les Johnson) would like to thank Stuart and Dolores Peck for allowing me to use their spare room so that I could have a quiet place to write on Wednesday nights. I also appreciate the constant supply of ice cream they provided. A person could not have better in-laws!

Some of the chapter frontispieces utilize C Bangs's photographs of exhibits in the Hall of Human Evolution at the American Museum of Natural History in New York City.

Synopsis and Chapter Summaries

Synopsis

This book describes a scenario for the re-greening of planet Earth. The book begins with a description of what our planet was like before the advent of our modern civilization, followed by a summary of the effects of that civilization on the planet, culminating with a discussion of how we humans might use the resources of the solar system for terrestrial benefit, resulting in Earth becoming a place for a technologically advanced human civilization to live in synch, if not in harmony, with the environment that gave us birth.

Even if human population peaks at 15 billion or so, utilization of extraterrestrial resources can reduce the stress on the terrestrial ecosystem. If extraterrestrial energy sources can be tapped and certain polluting industries moved from Earth's surface to near-Earth space, a substantial fraction of humanity can ultimately enjoy the lifestyle of modern-day Americans, Europeans, and Japanese.

This book builds on the success of the Matloff, Johnson, and Bangs book, *Living Off the Land in Space* (New York: Springer-Copernicus, 2007). It is an updated successor to the popular book *The High Frontier* (New York: Morrow, 1977), authored by Prof. Gerard K. O'Neill of Princeton.

Paradise Regained is intended for a wide audience. Authors Johnson and Matloff contribute their technical expertise and writing experience. The chapter frontispiece art of artist Bangs adds a significant visual dimension to this proposed solution to our environmental dilemma.

Chapter Summaries

Introduction: Welcome to the Present

Chapter 1: Space Utilization: A Moral Imperative

It is in this chapter that we will put forth a moral concept on which this book is based. It is a concept that should have nearly universal appeal and should guide much of our decision making regarding both space and environmental policies. Simply stated, life is good. The converse is also a moral assertion: that which leads to non-life is evil. Those who seek to preserve life, human and nonhuman, are acting in a morally superior manner compared to those who seek to diminish or harm life. We believe the moral decision that life on Earth is good drives those in the modern environmental movement to their activism. It is this same moral decision that motivates many space enthusiasts, activists, and professionals. We discuss why space industrialization and utilization is a viable, long-term, moral solution to our environmental problems.

Part I: Mythical Paradise

In this section we describe the formation and resources of the solar system and the ecological state of Earth before humans.

Chapter 2: Fire: Formation of Earth and the Solar System

This chapter summarizes current theories regarding the formation of the sun and the solar system. Particular emphasis is placed on planetary formation and the composition of the asteroids and comets. (This information is referenced in later chapters as we discuss how these resources might be used for supporting our future technological needs.)

Chapter 3: Earth Before Man: Utopia or Nightmare?

In the infant solar system, many comet-like bodies drifted through the inner solar system. Over time, comet impacts increased infant Earth's inventory of water vapor. As the planet cooled, this material condensed to the liquid state and Earth's oceans were born. According to current theories, the oceans were incubators for early life. Life emerged from these oceans to begin colonizing Earth's land surface. For billions of years, life evolved and diversified on planet Earth. Starting with single-cell bacteria, algae, and protozoa, multicelled organisms developed and the world they created was one of cold-blooded survival of the fittest. Without sentience, there was no concept of love, justice, or caring for those that are ill or weak, and certainly no concept of art, music, mathematics, or literature. Is this "pristine" Earth a viable model upon which we should base our environmental policies so as to re-create it?

Part II: Paradise Lost?
In this section we describe the environmental challenges facing humanity and describe why conservation and Earth-based alternative technologies will not be sufficient to avoid widespread ecological disaster—or the ultimate collapse of our technological civilization.

Chapter 4: The Environmental Dilemma: Progress or Collapse?
The second law of thermodynamics tells us that no energy conversion process is 100 percent efficient. There is a corollary to real-world engineering processes; no matter how hard we try, we will always have losses that will keep us from recycling with 100 percent efficiency. As demand grows and the supply of raw materials inevitably dwindles, recycling will become more essential, but in the long term, it, too, will fail to meet our needs. We will simply run out of resources. Long before that time, the pollution from a burgeoning materialistic (and prosperous) society will strain the ability of planet Earth to cleanse itself and we will run the risk of extinction from environmental degradation. The crisis may not be realized until 50 years from now or until 500 years from now—but it will happen. This chapter summarizes the challenges facing our civilization and our planet and asks the question, Do we want continued growth and progress, or collapse? It is in this chapter that we offer a potential solution to these problems through the development of space.

Chapter 5: Exploding Population
This chapter traces human population growth throughout history and ties it to the fraction of the species that today lives in relative prosperity due to their reaping the benefits of being part of a modern technological civilization. Past theories of population growth, forecasts of population collapse, and how we have continued to find technological paths to avoid such collapse are discussed.

Chapter 6: Climate Change
Putting aside the issue of whether or not the observed changes in the global climate are man-made, it is clear that our civilization still faces a tremendous challenge in dealing with observed and predicted changes in Earth's climate. For example, if the polar icecaps melt, then what impact will rising sea levels have on our civilization and economies? The predicted effects of climate change are outlined in this chapter.

Chapter 7: Vanishing Life

Planet Earth as a place for life is in jeopardy. Expanding human population, combined with climate change, is straining the sustainability of the few remaining habitats for wild life. Recent studies show that a distinct species of plant or animal becomes extinct approximately every 20 minutes; it is therefore conceivable that only our pets, food animals, and other domesticated species will survive the onslaught. We are poisoning our nest and killing, to the point of extinction, many species without malice of forethought, but by their misfortune of being in the way of our progress. This chapter provides a snapshot of the problem and discusses what experts in the field are predicting for the years to come.

Chapter 8: Diminishing Energy

The world's economic growth is driving an ever-increasing need for energy. Current methods of producing power are generally not environmentally friendly, and those that are green are often politically unacceptable to populations near where they are best implemented. This chapter discusses the recent history of energy production, various methods by which power is currently produced, and the environmental impacts of each. Specific technologies to be discussed include fission power, fusion power, coal, oil, natural gas, biofuels, wind power, and terrestrial solar power generation.

Chapter 9: Humans Before the Industrial Age: A Desirable Ecological Goal?

A common myth in popular culture is that we need to enact environmental policies that will return Earth to the near-pristine state of preindustrial human civilization. While this goal for planet Earth may be laudable, it is hardly an acceptable future for humanity. Modern civilization provides a quality of life that is vastly superior to that of our ancestors and it is doubtful that most of humanity will aspire to return to significantly shorter life spans full of rampant disease, poverty, and social inequity. In this chapter we discuss the lives of the common person in preindustrial society and assert that today's quality of life is morally superior despite the ecological consequences, considering that something must be done to prevent our hard-won societal progress from being lost as a result of our lack of good stewardship of Earth.

Part III: Paradise Regained

In this section we describe how the development of space and its resources not only can avert environmental disaster, but also can provide the basis for continued technological and societal progress.

Chapter 10: Raw Materials from Space

Space is not empty. Rather, it is populated with large rocky and gaseous bodies called planets, and millions of smaller bodies composed of rock and water ice (asteroids and comets). These smaller bodies contain many of the raw materials needed for our industrial society: metals, carbon compounds, silicon, and sometimes water. It is possible not only to mine them for these resources, but also to divert them to Earth for easier access. This chapter describes the resources of the solar system, referencing them against our projected industrial needs, and how we might use them to feed our industry once terrestrial sources become depleted or inaccessible.

Chapter 11: Power from the Sun

Sunlight reaching Earth contains 1368 watts of power per square meter. Assuming only a modest increase in sunlight-to-electricity conversion efficiency, it is conceivable that large solar array "farms" in Earth orbit could collect gigawatts of power and beam it back to the terrestrial power grid in an environmentally responsible way. The idea is not new, but the technologies that might make it possible are rapidly advancing. This chapter describes the concepts envisioned for space-based power generation systems and how they will work synergistically with Earth-based conservation to result in plentiful energy for the world's growing needs.

Chapter 12: Environmental Monitoring from Space

Earth orbit provides an unparalleled vantage point for monitoring global environmental change. The current generation of space satellites is providing environmental scientists and policy makers with nearly real-time information regarding global atmospheric and sea temperatures, sea levels, rainfall rates, atmospheric trace gas composition and variability, as well as numerous other indicators. In this chapter we discuss current, planned, and potential future capabilities for monitoring Earth's weather and climate from space.

Chapter 13: Protecting Earth

From the asteroid impact that is thought to have made the dinosaurs extinct 65 million years ago to the rock that hit Tunguska, Siberia, in 1908 with the force of a 20-megaton nuclear weapon, Earth is in the shooting gallery and it is only a matter of time before we are impacted again. (The Tunguska event was about 1000 times more powerful than the atomic bomb dropped on Hiroshima, Japan.) Fortunately, we have the ability to go into space and divert any potential threats to the planet—a capability the dinosaurs did not have. This chapter discusses the impact threat, provides an overview of past impact events, and discusses the options we have for averting similar disasters by detecting, characterizing, and diverting any objects that might impact Earth.

Chapter 14: Mitigating Global Warming
According to the International Energy Agency, it is projected that carbon-dioxide emissions will more than double by the year 2050, with developing countries accounting for almost 70 percent of the increase. Even with draconian emission reductions on the part of the industrialized world, global CO_2 emissions will continue to grow at an alarming rate and, if current theories are true, the world will continue to get hotter. Several ideas have recently been proposed for mitigating this global temperature rise by blocking a small amount of the sunlight reaching Earth using space-based filters or sunshades located at one of the Earth/sun Lagrange points. These approaches are discussed in the context of how they might be used in conjunction with conservation and slowing the growth of greenhouse gas emissions on Earth to reduce the global temperatures to pre-20th-century levels.

Chapter 15: Settling the Solar System
To thrive on Earth in large numbers, humans must experiment with various forms of renewable energy and lifestyles that are less degrading of the environment. Many of these techniques are similar or identical to those that will be required to settle nearby regions of the solar system. How might we use solar energy and closed ecological systems in space to plant self-sufficient human habitats on the moon, on Mars, or in free space?

Chapter 16: Paradise Regained: An Optimistic Future
The marriage of environmental action and space technology implementa-tion can achieve what might otherwise be impossible—the return of Earth to being a planet for life (all life, not just our species) with an ever-increasing standard of living and quality of life for all its inhabitants. Space-based industries can produce many of our industrial products without "fouling the nest." Solar power satellites can augment terrestrial sources, obviating the need for more fossil and nuclear-fueled power plants. Asteroids can be diverted to avert global disaster and to provide nearly infinite resources. And Earth can once again be a place for life. This chapter summarizes how space technologies can work synergistically with conservation, recycling, and planned growth to build a prosperous and sustainable future for all of humanity.

Author Biographies

Les Johnson

Les Johnson is the deputy manager of NASA's Advanced Concepts Office at the George C. Marshall Space Flight Center in Huntsville, Alabama. Previously, he managed NASA's In-Space Propulsion Technology Project, developing advanced technologies such as solar sails and aerocapture for future space science missions. From 1996 through 2003 Johnson was the principal investigator for the ProSEDS tether propulsion experiment. In addition to his NASA credentials, Johnson also is an inventor. He holds three patents, was twice the recipient of NASA's Exceptional Achievement Medal, and is the author of numerous technical publications, coauthor of two mass-market popular science books, and has consulted on various novels and a major motion picture.

Gregory L. Matloff

With an M.S. in astronautical engineering and a Ph.D. in applied science (both from New York University), Greg Matloff has published or delivered about one hundred research papers related to atmospheric physics, space exploration, or space science, and has authored or coauthored seven books and many popular articles. As well as teaching physics at New York City College of Technology, he consults for NASA Marshall Space Flight Center, is a corresponding member of the International Academy of Astronautics, a fellow of the British Interplanetary Society, and a Hayden Associate at the American Museum of Natural History. With his wife C Bangs, he resides in Brooklyn, New York.

C Bangs

With a B.F.A. from University of the Arts and an M.F.A. from Pratt Institute in painting and sculpture, C Bangs has exhibited her art in museums and galleries throughout the United States, South America, Europe, and Australia. She has created the chapter frontispiece art for the books authored by her husband, Greg Matloff. Her work has appeared in the

Journal of the British Interplanetary Society, Analog: Science Fact and Fiction, and *Zenit.* She worked under a grant at NASA Marshall Space Flight Center and then as a NASA faculty fellow for three sequential summers. Her art is included in numerous public and private collections.

Introduction: Welcome to the Present

I know a bank where the wild thyme blows,
Where oxlips and the nodding violet grows;
Quite over-canopied with luscious woodbane,
With sweet musk-roses and with eglantine

William Shakespeare, *A Midsummer Night's Dream*

If you are reading this book, you are most likely a member of the most privileged generation in the history of humanity. You have a roof over your head—a vast improvement over the lot of many of our ancestors and a significant number of humans today. You have access to quality health care facilities and can count on between 70 and 80 good years of life.

Clean water is yours at the twist of a knob. And food—as healthful or exotic as you might desire—is available a short walk, subway ride, or drive from your front door. Your great-grandparents might have sacrificed a great deal for these advantages alone.

Education, although not universal, is widespread. In all likelihood, you have a moderately creative and financially rewarding professional life—also a great rarity before about 1950. If you push a button, entertainment and information from the world over can flood your consciousness. You can even remain connected to this planet-wide information network as you walk in a park, cook your dinner, or ride in a car.

But problems loom on your, and our, horizon—problems that threaten to swamp our glorious era. The world's population continues to expand. And surprisingly to some (but not to us), the burgeoning populations of Asia, Africa, and South America desire the same advantages enjoyed for many years by North Americans, Europeans, Japanese, and Australians.

Can contemporary civilization provide for billions more humans living well? Where will the energy come from? How do we deal with the pollution? Will carbon dioxide emissions and other human-produced greenhouse gases evoke long-lasting climate change that will increase global temperatures, raise sea levels, and swamp coastal lands?

Years ago, as our ancestors began to emerge in the park-like savannah of central Africa, a response developed to the problem of environmental degradation. When your local environment was exhausted, move! This worked well as hunter-gatherers spread around the globe and civilizations developed and spread. The world is littered with the ruins of once-great cities surrounded by degraded environments.

But today there is nowhere to flee. Civilized humans are everywhere on Earth. Early science-fiction authors hoped for benign climates on neighboring worlds, but none of these worlds could sustain more than a tiny fraction of the human population, and that at great expense. Even if people were genetically modified with gills to live in the oceans, this would be a mere stopgap measure; the oceans are not immune to human pollution.

Conservation, limiting growth, and recycling can provide some relief, but any such relief will almost certainly be only temporary. With the rest of the world's economies growing into variations of our materialistic one, it is only a matter of time before we simply run out of resources, energy, and places to store our waste. This does not mean we should not conserve, limit growth, and recycle! On the contrary, these measures are essential to the survival of our civilization and, potentially, our species. They are simply insufficient to resolve the core issue we face—that Earth cannot by itself indefinitely sustain a worldwide population of consumers. It is impossible to recycle with 100 percent efficiency, to mandate no growth, or to conserve our way into prosperity.

It was a bitter day in December 1968 when humans in general became aware of their kinship as riders on a fragile, living Earth. From a quarter million miles away, the crew of Apollo 8 pointed their cameras homeward after they had safely settled into humanity's first close flyby of the moon.

The view of the desolate moon was striking on our television screens, and the astronauts' scripted reading from Genesis was stirring. But it was the shimmering, living, marble-sized Earth, hanging in stark contrast above the lunar horizon that would profoundly alter our view of our world and ourselves.

Living worlds are fragile and delicate. And space is very, very large. This lesson would be repeated and amplified in 1990, when, on the edge of the galactic void, the cameras of Voyager 1 were focused back on our planet. From its multi-billion-mile vantage point, Voyager imaged Earth as a pale, blue dot almost lost in the glare of the distant sun. Earth is a seemingly unique abode for life in an otherwise empty and apparently lifeless expanse of nothingness.

So as you escape our increasingly urbanized world to stroll through your local park, botanical garden, or forest, and gain spiritual sustenance from your temporary immersion in this sanitized (predator-free) version of our original environment, you might well ponder the troubled legacy of our golden age. Planets abound in our galaxy, but planetary ecospheres are rare and precious.

Will the parks of Earth survive this age of overpopulation, resource consumption, nuclear proliferation, and terrorism? Or will our civilization go down with a bang or a whimper, to be remembered in legend by a remnant population eking out a living in a depleted, contaminated landscape?

No one can know the answer to this question. But there is hope within the gloom. An optimistic scenario exists if contemporary humans are collectively wise enough to grasp the opportunities of the present. While the modern environmental movement applies much-needed first aid to our resource and environmental challenges, we should simply look up for the potential cure; space harbors enough resources to meet the needs of an ever more prosperous humanity for millennia. Space is an environment already hostile to life that can be used to house industries whose by-products are also antithetical to life, and it may provide nearly infinite energy.

Our solar system is very rich in energy and resources and can even serve as a sink for some of the unavoidable effluents of technological society. To avoid the fate of the dinosaurs, humans may decide to use our revived interplanetary capabilities to alter the solar orbits of those kilometer-sized chunks of cosmic rock and ice that occasionally wallop Earth. And if we can move these objects around, perhaps we can mine some of them.

Even if interplanetary space can never absorb more than a handful of Earth's human inhabitants, its resources can be used for the betterment of life. Envision a future where plentiful energy comes from the sun, industrial pollution is virtually removed from the ecosphere, and no country need suffer from a lack of natural resources. Then, the 10 to 15 billion peak human population on planet Earth can enjoy comfortable, productive lives. And the parks of Earth, the thyme and violets, need not die.

This book is divided into three parts. Part I, Mythical Paradise, reviews current scientific thinking about how Earth came to exist and how life arose. Many consider primitive Earth, the planet as it was before the rise of human civilization, to be a paradise lost. In reality, it was a hostile environment in which only the strongest survived, and it was far from being a paradise.

Part II, Paradise Lost?, describes the rise of human civilization, our progress from simple daily survival to where we are today—with many humans living long, productive, and meaningful lives—and the associated (mostly negative) impacts to the environment that our civilization has wrought. It is here that we outline the challenges our 21st-century civilization faces.

Part III, *Paradise Regained*, describes how space and space technologies can be used to monitor the global environment, help undo ecological damage, and prevent further damage to Earth's ecology upon which we all depend.

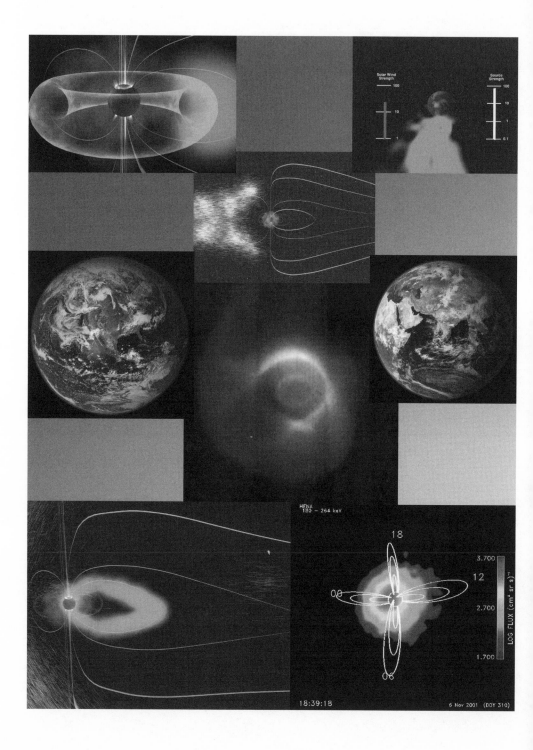

Space Utilization: A Moral Imperative

> I saw the sudden sky;
> Cities in crumbling sand;
> The stars fall wheeling by;
> The lion roaring stand.

W.J. Turner, from *The Lion*

Philosophers and theologians have debated what constitutes morality since there have been people around to consider the notion. The word *morality* comes from Latin and refers to our notion of what constitutes right and wrong or good and evil. What constitutes a moral action varies dramatically from culture to culture, though many cultures share some common thought as to what is right and wrong. For example, most cultures believe that murder is not a morally acceptable method to settle a dispute and that theft is not a moral act. But even these seemingly simple moral judgments are far from universal, and anthropologists can undoubtedly find cultures on the globe that do not share these views.

We will put forth a moral concept upon which this book will stand. It is a concept that should have nearly universal appeal and should guide much of our decision making regarding both space and environmental policies. The moral statement that drives our thesis is deceptively simple. In fact, it is so simple that an entire company is built around it as both a trademark and a pithy statement: Life is Good™. The converse is also a moral assertion—that which leads to non-life is evil. And by "life" we mean communal life, not that of an individual animal or plant, although in the case of humans we believe the assertion is valid almost 100 percent of the time. Those who seek to preserve life, human and nonhuman, are acting in a morally superior manner compared to those who seek to diminish or harm life. Often we must act on this moral principle in such a way that some life is harmed along the way to a more global solution that greatly preserves or improves the quality of life in general. These choices must be carefully considered in the context of whether or not they will improve the quality of life or maintain life in the "big picture" or on a scale beyond the immediate and obvious impacts to an individual.

L. Johnson et al., *Paradise Regained*, DOI 10.1007/978-0-387-79986-5_1,
© Praxis Publishing Ltd, 2010

We believe the moral decision that life on Earth is good drives those in the modern environmental movement to their activism. It is this same moral decision that motivates many space enthusiasts, activists, and professionals.

A moral declaration is that life is good and better than non-life. Once this declaration is made, it is easy to see that in order for humans to prosper and be good stewards of the planet and its myriad life forms, they must stop ruining the environment that gives it life. But how do we accomplish this goal and maintain all of the positive aspects that come from our 21st-century technological civilization? The answer is simple. We must place heavy industry, with all of its inevitable pollution, in a place that is already hostile to life and in which life will almost certainly never arise—in space or on the moon. A space-based infrastructure centered on Earth/moon space is not only possible but also essential if we are to return our home planet to being a place to live, not a place to pollute. This chapter explores the current environment of near-Earth space, so as to address concerns by some that our presence there will somehow pollute it.

Most of us think of outer space as being empty. If we are comparing space starting at about 200 kilometers (km) above our heads and extending infinitely outward, to what it is like around us on Earth, then *empty* seems to be the appropriate word. For all practical purposes, to most humans, space is empty. (When we refer to space in this context, we are referring to that which is separate and apart from a planet, comet, or asteroid. Of course, the total volume of space occupied by planets, comets, and asteroids is so small in comparison to the rest that it is inconsequential overall.) In space, there is no atmosphere to breathe, no potable water to drink, and not much else in significant quantity—at least from the average human being's perspective.

However, in the 50 years of space exploration, scientists have learned that space is far from empty, and most of what is found in space is directly antithetical to life. To begin with, the entire solar system is bathed in sunlight. We depend on the light emitted by the sun to sustain us. But this light is only a small part of the total electromagnetic radiation ("light") emitted by the sun. We see only this part because our dense atmosphere filters out most of the rest. The sun emits light at several wavelengths, from infrared to extreme ultraviolet. The part of the sun's spectrum that we see is only a very small part of the emitted spectrum and is aptly called "visible light" (Fig. 1.1). By definition, the other parts of the spectrum are invisible to our eyes, but not to our scientific instruments.

Over the past several years, there has been much discussion about solar ultraviolet light. It is this part of the spectrum that causes people to tan and sunburn—in many cases, subsequently causing skin cancers. Fortunately

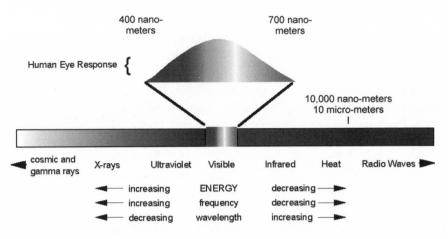

Figure 1.1: The sun emits light at many wavelengths, only a small fraction of which can be seen by the human eye. This portion of the spectrum is known as "visible light." (Courtesy of the Universities Space Research Association.)

for us, most of the ultraviolet light from the sun is filtered at high altitudes by atmospheric ozone. (There are multiple wavelengths of light that are considered to be part of the ultraviolet spectrum. They are divided into two parts: ultraviolet A [UVA] and ultraviolet B [UVB].) The amount of UVB light that penetrates through the atmospheric ozone decreases rapidly with increased ozone concentrations. The converse, unfortunately, also applies: decreased ozone will dramatically increase the amount of UVB reaching the surface of Earth. UVA passes roughly unhindered through the atmosphere.

Since the 1980s, and mostly likely beginning earlier, a global decrease in atmospheric ozone density was observed, with especially large depletions occurring near Earth's poles. The large depleted regions near the poles were dubbed "ozone holes," and their existence was blamed on the human emission of ozone-depleting chemicals such as chlorofluorocarbons. This depletion was of concern because increased UVB exposure would increase the number of people getting various cancers. It is also thought that increased UVB would do significant harm to plankton in the oceans, with harmful ripple effects felt throughout the food chain that depends on plankton. Also, several crop species are thought to be UVB sensitive, with death or significantly reduced yields resulting from increased exposure. Global environmental action was taken in the 1990s to reduce human emission of ozone-depleting chemicals so as to mitigate further ozone depletion.

Solar ultraviolet radiation acts as a sterilizer and quickly kills any unprotected life. In fact, ultraviolet sterilization is used commercially to kill bacteria in swimming pools and in air purification systems. In space, there is no atmospheric ozone to filter any of the UVB, and ultraviolet radiation from the sun is deadly.

In addition to visible and ultraviolet light, the sun emits many other forms of radiation. High-energy electrons and protons continuously stream from the sun; this stream is commonly called the solar wind. The sun emits these particles with velocities between 200 km/sec and 800 km/sec, depending on where you are relative to the sun's equator and the solar activity cycle. When these charged particles encounter matter, such as living tissue, they deposit their energy as they slow down and stop therein. These particles have a lot of energy, and the slowing-down process does significant damage to any living tissue in which it occurs. Even short-term exposure to unfiltered sunlight can cause mutations and cancer. Moderate to long-term exposure results in significant cell damage and death.

And it gets worse. In addition to this somewhat steady stream of charged particles coming from the sun, there are periodic bursts of intense high-energy radiation, called solar energetic particle events, that blast lethal storms of charged particles outward from the sun to the outermost regions of the solar system, each packing enough energy to cause human death within minutes to hours of exposure.

Fortunately, our Earth once again protects us from this danger. Earth's relatively strong magnetic field acts as a radiation shield, deflecting all but the most energetic of these particles away from the surface of the planet. (Charged particles, when moving through a magnetic field, experience a force acting on them, in this case, a deflecting force.) Our thick atmosphere, which has the approximate stopping power of 10 meters of water or 4 meters of concrete, absorbs most of the remainder (Fig. 1.2).

On Earth, we go blithely through our days during these storms, blissfully unaware of the hellish inferno blasting through space a mere few hundred kilometers above our heads.

Across the globe, the average temperature does not vary by much, providing a reasonably stable thermal environment for all sorts of life. Where seasonal variation does occur, life, in general, has adapted to it, growing during the warmer summer months, hibernating during the colder winter months. In author Les Johnson's adopted hometown of Huntsville, Alabama, the average temperatures range from a low of about 29° Fahrenheit in January to a high of 89° in July. Most residents of north Alabama adapt quite nicely to this range of 60 degrees. Thanks to the complex biosphere of Earth, with enormous oceans and a thick atmosphere to average out and regulate thermal effects, 60-degree temperature variations in the course of a

Figure 1.2: Earth's magnetosphere, depicted here, acts as a shield against all but the most energetic cosmic rays, preventing them from reaching the surface of Earth in great quantity. (Courtesy of the National Aeronautics and Space Administration [NASA].)

year are typical for Earth's nontropical, nonpolar regions, which are typically mild and for which life has readily adapted.

In space near Earth, without an atmosphere or ocean, the temperature can vary from +200° Fahrenheit to –200° as quickly as an object moves from being in sunshine to being in darkness. Most forms of life would either freeze to death or be cooked in these 400-degree temperature swings.

If you go out on a clear night in the middle of August, you have a pretty good chance of seeing a shooting star—more accurately called a meteor. If you know where and when to look, you will likely see several. It is during this time that the Perseid meteor shower occurs, a result of Earth's passing through the trail of pea-sized debris left behind by comet Swift-Tuttle. Harmless to us on the surface of the planet, these small pieces of cosmic dust and ice hit the atmosphere at 60 km/sec and quickly burn up, leaving a glowing trail across the night sky. The Perseids are a harmless, beautiful sight, unless you happen to be above the protection of our dense atmosphere.

If you are in space or on the moon, these pebbles traveling at 60 km/sec are more deadly than a speeding bullet (which travels at only about 0.8 km/sec). When being hit by a bullet, a fist, or a car, what matters is not the overall speed, but the kinetic energy of the impactor. For the most part, the energy transfer from the impactor to you determines how much damage you sustain. Kinetic energy is the energy associated with the motion of a body, and the amount of kinetic energy depends on both the mass of the object and the *square of its velocity*. In practical terms, this means that you can double the amount of damage done by doubling the mass of the projectile and simply keeping its speed constant. However, if you double its speed, you do not double its kinetic energy, you increase it by a factor of 4. A bullet shot from a gun on Earth will have only a fraction (about 0.017%) of the energy of that same bullet moving with the speed of a Perseid meteor. It will therefore do only a fraction of the damage caused by being at the wrong place at the wrong time and getting hit by a meteor!

Granted, the probability of getting hit at any given time is low. There are only so many meteors, and space is very big. But once you factor in the frequency and damage potential of the larger impactors such as asteroids, the risk begins to rise sharply. Getting hit with a large rock moving at tens of kilometers per second is like having a hydrogen bomb dropped on you, or worse. (See Chapter 13, in which we discuss the threat from near-Earth objects and how one object impacting Earth is thought to have ended the reign of the dinosaurs.) The bottom line is that we live in a cosmic shooting gallery, and getting hit by any of these objects would be the perfect definition of a bad day.

One aspect of space that is foremost in most people's minds when they think of life in space is vacuum. What happens to living things in the vacuum of space? This is an easy question to answer. Unlike depictions in many science-fiction movies, the human body does not inflate like a balloon and explode when exposed to vacuum, nor does the blood boil or immediately freeze. You may not even immediately lose consciousness. But unless you get back under pressure very quickly, you will most assuredly die—from asphyxia.

If you were to step out on the surface of the moon and somehow survive the vacuum, you would still have to worry about the continuous exposure to solar radiation in all its forms, such as visible light, ultraviolet light, and infrared light, as well as the charged particles in the solar wind and those resulting from solar energetic particle events. This radiation can quickly reach levels that even a moderately protected human could not survive. Next, you would have to protect yourself from the extreme heat during the 14-day lunar day and the extreme cold of the 14-day lunar night. (The moon

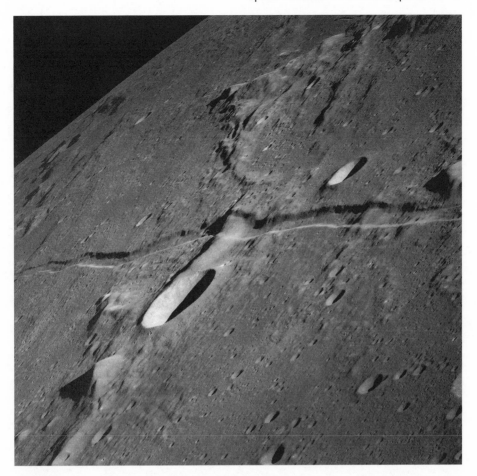

Figure 1.3: The moon is a dead environment without life—until we go there and build our cities and factories. (Courtesy of NASA.)

rotates more slowly on its axis than does Earth. One lunar day is therefore equal in length to 14 Earth days. We see only one side of the moon because its rotation rate corresponds to the rate of lunar revolution about Earth.) If you are standing on the moon, you are standing on what may be the deadest place in the nearby solar system. The moon is blasted by intense solar radiation, remains in vacuum, and is alternatingly cooked and frozen on a regular basis. There is probably not a deader environment anywhere close by in the solar system and certainly not anywhere on Earth (Fig. 1.3).

Without human intervention, space is anti-life. With the possible exception of the bacterium *Deinococcus radiodurans,* no life as we know it can survive in this hostile environment without the artificial protection such as that which we humans devise for our space explorers. Given that there is no air or water to pollute, no ozone to deplete, no climate to change, no environment to harm with the radioactive by-products of our nuclear power plants (the radiation from the sun makes our pitiful human radioactive output seem truly minuscule by comparison), and no ecosystem to clutter with our waste products, what can we humans possibly do in space to make it more anti-life than it already is? Absolutely nothing.

In fact, the argument can be made that by expanding the realm of human activity to space, including all the processes and products that on Earth would be called pollution and pollutants, we will be creating new places for life to exist and thrive. Such expansion would be a thoroughly positive moral choice. Our industrial plants will have to have breathable air and drinkable water, they will have to have artificial protection from solar radiation in all its forms, and they will have to regulate the temperature so that human life can survive and thrive. We will be creating "green" ecosystems from desert, and the inevitable by-products of our civilization, the pollutants, will not harm any ecosystem in any way. We should not be profligate and wasteful by any means. Our explorers and industrialists will not want to waste anything that has potential use, because it simply will be too expensive to do so. Recycling should be the norm and only after all other options are exhausted should we discard our waste into the space environment.

We have a moral obligation to develop space resources and to foster space industrialization. To not do so is ultimately anti-life and an immoral act of omission.

Mythical Paradise

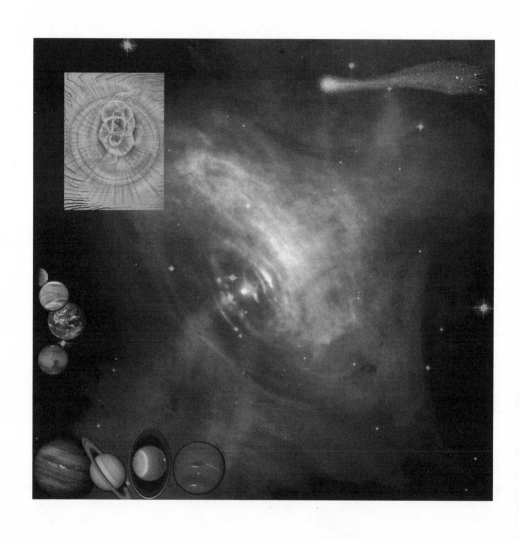

Fire: Formation of Earth and the Solar System

It lies in Heaven, across the flood
Of ether, as a bridge
Beneath the tides of day and night
With flame and darkness ridge
The void, as low as where this earth
Spins like a fretful midge.

Dante Gabriel Rossetti, from *The Blessed Damozel*

In the beginning, a cloud composed mainly of hydrogen and helium gas drifted through the interstellar void. Near or within this immense nebula (it must have been trillions of miles across), a bright star blazed. Much larger and more massive than our present-day sun, this nameless star approached the end of its life cycle about 5 billion years ago. As its nuclear fires waned, it began to collapse. As it collapsed, temperature and density near its core increased dramatically.

Suddenly, conditions in the stricken star's interior became sufficient to support nuclear reactions more elaborate than the hydrogen fusion cycles that support our sun and most other mature stars. In a spectacular explosion called a supernova, elements as massive as uranium were bred.

For a short while, the dying star outshone the entire stellar host in our Milky Way galaxy combined. Before its embers faded to obscurity, the heavy-element-laden gases it emitted began to mix with the lighter elements in the neighboring nebula.

As remnants of the exploded star diffused through the giant nebula, turbulence developed. In slow motion (at least from the human viewpoint), eddies and whirlpools began to develop throughout the vast celestial cloud. The largest of the eddies, condensing under self-gravitation, would eventually become stars. Smaller eddies would someday result in planets, icy comets, and rocky asteroids.

The nebula had now become a star nursery. Through the telescopes of any advanced extraterrestrial civilization in that distant era, the nebula containing the infant sun and solar system (and many others) may have

L. Johnson et al., *Paradise Regained*, DOI 10.1007/978-0-387-79986-5_2,

resembled Messier 42,* the great nebula visible through binoculars below Orion's belt in the northern skies of spring.

Eons passed. Material gathered in the large proto-star eddies and in the smaller proto-planets. Continent-sized fragments of rock and ice careened through the infant solar systems, occasionally colliding with a proto-planet or being gobbled up by a proto-star.

Although little has survived on Earth from this ancient era, asteroids and comets are remnants of the solar system's formative phase. Our telescopes on Earth and in space offer tantalizing glimpses of star formation still occurring throughout our Milky Way galaxy.

As matter accumulated around the proto-sun, the temperature and density near this object's interior gradually increased and suddenly were sufficient to support a thermonuclear reaction dubbed "hydrogen burning." A hydrogen-burning star, such as our sun, emits light through this process. Deep within the stellar interior, hydrogen nuclei are fused together to produce helium, energy, and mysterious neutrinos.

The neutrinos, which have little or no mass, are particles that are nonreactive with normal matter. They rapidly traverse the star's interior layers and disappear in the depths of space, carrying with them a significant fraction of the fusion reaction's energy output. Most of the energy released in the stellar interior is a type of invisible, high-energy light called gamma rays. By the time this energy has reached the star's outer layer (called the photosphere), much of it is in the form of light visible to our eyes.

After millions of years of contraction, the sun had finally turned on! First, the pressure of the solar radiation slowed and stopped the contraction. Light from the infant sun streamed through the turbulent new solar system. The stream of magnetically accelerated high-energy electrically charged particles called the "solar wind" started up and soon began to influence the nebula regions closest to the infant star.

Near the Central Fire

About 4.7 billion years ago, the planets began to take form. Although the planets initially were mostly hydrogen and helium, as is the sun, the solar

---

* The French astronomer Charles Messier in the 1700's cataloged numerous objects in the night sky that were not planets, stars or comets with a series of numbers beginning with the capital letter M. He did not know what these objects were. The Orion Nebula is an interstellar cloud of dust, hydrogen gas and plasma from which new stars are being born.

wind drove off the initial cloud of light gas enshrouding each of the inner planets.

Close to the sun, tiny moon-like Mercury coalesced from the dust, gas, and rock in the inner solar system. With too little mass to long maintain an atmosphere or oceans and much too hot, this tiny world is forever barren. But comets in large numbers traversed the young solar system, and some of them impacted Mercury, apparently leaving frozen water deposits in sun-shielded craters near the planet's poles.

Farther from the sun is the planet Venus. Since this namesake of the ancient love goddess is a near twin of Earth in terms of size and mass, early astronomers hoped that swamps or oceans might exist beneath the cloud banks of this world. But alas, probes from Earth have penetrated the clouds to learn of this planet's true conditions. Temperatures at the surface are hot enough to melt lead, and atmospheric pressure at the surface of Venus is about ninety times that at Earth's surface. In addition, the atmosphere of Venus is mostly carbon dioxide—a waste product of terrestrial animal life. Even worse, a steady rain of highly corrosive sulfuric acid drips from the leaden skies over Venus.

Some love goddess! Even the best-shielded probes from the United States and from Russia have survived no more than a few hours on the surface of Venus. Although it is a lovely sight in the evening or pre-dawn sky, Venus is about as close to the medieval concept of hell as any world we are likely to discover. It is unlikely that humans will ever visit this hothouse world, let alone live on it.

A number of theories have been proposed to explain what went wrong on Venus. One likely possibility is that comets and asteroids near Venus are closer to the sun than they are in Earth's vicinity and consequently move faster. When such objects impact Venus, their high kinetic energy relative to that planet is converted into heat.

Combined with the higher ambient temperatures on Venus (since its separation from the sun is about 70 percent of the Earth–sun separation), conditions were such that the oceans caused by impacting comets evaporated almost immediately. Venus was shrouded from the start by an atmospheric envelope of carbon dioxide. This gas absorbs infrared radiation reemitted by the planet, causing the runaway greenhouse effect that has precluded the evolution of any life we could imagine beneath the cloudy veil of the love goddess.

The next planet out, about 93 million miles from the sun, is Earth. We will return to Earth after we survey the formation of other solar system bodies in the early solar system.

About 50 percent further than Earth from the sun is the red planet, Mars.

Perhaps because of its blood-red color, this planet is named after the ancient war god. Although Mars has a day only slightly longer than that of Earth, it has only 10 percent of Earth's mass. Like Earth, Mars was bombarded early in its existence by water-bearing comets. It also has volcanic mountains larger than any on Earth. In spite of these sources of volatiles and its distance from the sun, Mars's atmosphere is far thinner than that of Earth. The planet's small size probably caused most of its atmosphere to escape into space.

Unlike Venus, Mars still has water reserves. There is some water vapor in the thin atmosphere and frozen in the polar caps. Recent observations by Mars-orbiting probes and rovers indicate that some water may also exist in the upper soil layers of the red planet. Although Mars may have developed life early in its history, living Martians appear to be very sparse or absent.

The Outer Realm

Although chunks of rock and ice routinely traversed the early solar system, collisions with planets and satellites has gradually cleared most inter-planetary space. Beyond Mars, we find the asteroid belt. Although some mountain-sized space boulders are found elsewhere in the solar system, most of the surviving rocky debris from the solar system's origin is now found in this realm. Some asteroids are as big as the state of Texas and are classified as dwarf planets.

Sometimes, because of collisions or orbital perturbations, objects from this asteroid belt reach Earth. If they survive the fiery descent through Earth's atmosphere, they are called meteorites. Meteorites are categorized in three basic classes: rocky, stony, and carbonaceous. Most rocky and stony meteorites originate in the asteroid belt. Water-rich carbonaceous meteorites may have their origin further afield among the comets.

Beginning about five times Earth's distance from the sun is the realm of the giant worlds: Jupiter, Saturn, Uranus, and Neptune. These worlds are rich in gases including hydrogen and helium, probably because sunlight and the solar wind are too weak at these solar distances to have evaporated them into space.

The largest of them, Jupiter, has about 318 times as much mass as our Earth, but only about 1/1000 the mass of the sun. Composed mostly of hydrogen and helium, Jupiter is often referred to as a star that failed. If this giant world were somewhat more massive, thermonuclear fusion would have ignited in its interior. We would then live in a double-star system!

But even as a planet, Jupiter is impressive. It is equipped with colorful cloud bands and an atmospheric disturbance larger than the Earth called the "great red spot." Like the four largest of its many satellites (Callisto, Ganymede, Europa, and Io), these features are easily viewed through binoculars or a small telescope. These large moons of Jupiter are worlds in their own right. Io features many active volcanoes; a tantalizing, mostly frozen deep-water ocean covers Europa. Many biologists expect to find life in the liquid regions of this satellite-wide sea. Although the largest four Jovian satellites likely coalesced in position like a miniature solar system during the dawn of our planetary system, most of the smaller satellites of this huge world are captured asteroids or comets.

As is true for all of the giant worlds, Jupiter is equipped with an encircling ring. Planetary rings originate either from the disintegration of satellites that approach the giant world too closely or from debris from a satellite that could not form due to the proximity of the planet. After the sun, Jupiter is the most intense natural radiofrequency source in the solar system. If you tune a short-wave radio to 20.10 MHz, much of the static you hear originated from Jupiter! Jupiter has a strong magnetic field and radiation belts that might pose a problem to future human visitors.

The average separation between Jupiter and the sun is about five times the Earth–sun separation. Almost twice as far from the sun is the next major planet, Saturn. In terms of mass, Saturn would seem outclassed by Jupiter, since it is less than one-third the mass of that planet. Like Jupiter, Saturn has many satellites. One of them, Titan, is the only planetary satellite in our solar system with a dense atmosphere. Although present-day terrestrial life would have a hard time breathing Titan's nitrogen-argon-methane atmosphere or swimming in its hydrocarbon seas, conditions on this fascinating world may be similar to those on primitive Earth (although a lot colder). Through binoculars or a small telescope, Saturn is an elegant sight. Its dramatic rings are spectacular under most viewing conditions.

Further out from the sun, in the frigid wastes of deep space, are the final two gas giants of our solar system: Uranus and Neptune. Methane and ammonia are major atmospheric constituents of these worlds, which are respectively about fifteen times and seventeen times the mass of Earth.

The rotation axis of most planets is usually fairly close to perpendicular to the direction of the planet's solar orbit. Not so for Uranus! Early in the solar system's history, something huge must have smacked this planet, knocking it on its side so that its rotational axis is nearly parallel to the direction of its solar orbit. This cosmic collision would have been something to behold!

Beyond the Planets

Beginning at the orbit of Neptune (about thirty times farther from the sun than is Earth) and continuing outward about the same distance, we encounter the icy objects of the Kuiper Belt. Called Kuiper Belt objects (KBOs), these dwarf worlds are composed mostly of frozen water, ammonia, and methane. When a collision or gravitational perturbation pushes a KBO sunward, some of this icy material melts. Heated by sunlight and affected by the solar wind, this material streams from the tiny, rocky nucleus of the KBO. It appears in our sky during these sunward passes as a short-period comet (meaning that its solar orbital period is less than 200 years).

The second largest known KBO, Pluto, was originally classified as the solar system's ninth major planet. Since the discovery of Eris, a KBO more massive than Pluto, the largest KBOs are now classified as dwarf planets. Most known KBOs are about 100 km in radius. A few like Pluto are about 1000 km in radius. Assuming the existence of many smaller objects in this region, there may be as many as 70,000 members of the Kuiper Belt.

Beyond the Kuiper belt is the Oort cloud, a much larger comet repository that originated with the solar system. Perhaps a trillion comets, each more than 10 km in radius, inhabit this annulus, which stretches to perhaps 100,000 times the Earth–sun separation. Sometimes a passing star nudges one or more of these frigid worldlets sunward. It then appears briefly in our skies as a long-period comet. Some of these have orbital periods of 100,000 years or longer.

Meanwhile, Back on Earth

As our planet coalesced in the early days of the solar system, it was certainly not a park; a picnic would have been no fun on infant Earth. This was no place to safely raise your child, brush your cat, or walk your dog. Continent-sized chunks of ice and rock, such as the comet shown in Figure 2.1, careened through the inner solar system, and many collided with the Earth. It must have been quite a show.

Each of these collisions was a double-edged sword. Huge craters were gouged in the molten crust of the young world; tectonic forces responded by exploding volcanic geysers into Earth's new sky. Any life that had taken hold on the new planet would be quickly extinguished by these catastrophic events. But at the same time, each collision brought new material from space. From comet impacts, Earth's primeval carbon-dioxide–rich atmosphere

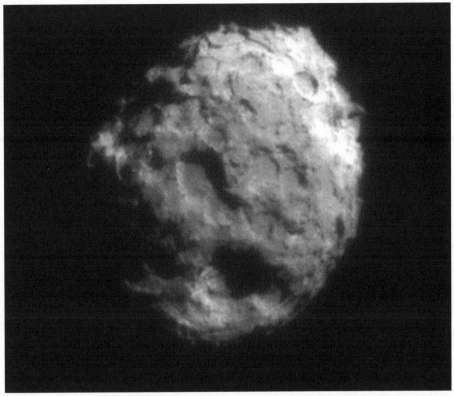

Figure 2.1: The nucleus of Comet Wild 2 as photographed by NASA's Stardust Mission. (Courtesy of NASA.)

began to form. Comet-supplied water fell from the skies in huge quantities, only to be evaporated by the high-temperature crust.

Over hundreds of millions of years, Earth gradually cooled. But about 4.2 billion years ago, Earth may have experienced the largest impact of them all. A world about the size of Mars, with about 10 percent the mass of Earth, is thought to have broadsided our planet. Once again, Earth was in a molten state. Much of the debris was flung into space, where it ultimately coalesced under self-gravitation to become our moon.

The population of comets and asteroids in the inner solar system began to thin. Earth's crust was, at least in selected regions, solid rather than molten. Finally, all of the water that fell from the skies did not immediately evaporate.

About 3.8 billion years ago, life began. We do not yet know the entire story of this remarkable development. But life had arrived on the young planet. Earth, while not yet a park, was no longer barren.

Further Reading

Solar-system data are available from a wide variety of sources. One of our favorites is K. Lodders and B. Fegley, Jr., *The Planetary Scientists' Companion* (New York: Oxford University Press, 1998).

Is Pluto a major planet or not? The popular astronomy press has done a good job of covering this debate. See, for example, F. Reddy, "Top 10 Astronomy Stories of 2006," *Astronomy*, 2006;35(1):34–43.

Earth Before Man: Utopia or Nightmare?

O World! O Life! O time!
On whose last steps I climb,
Trembling at that where I had stood before;
When will return the glory of your prime?
No more—Oh never more!

Percy Bysshe Shelley, from "A Lament"

There can be no question that today, one species, *Homo sapiens,* bestrides the world. In some circles, it is fashionable to lament this situation. Has a Golden Age been lost; has Eden been transformed into an omnipresent global civilization of commerce, popular culture, consumerism, and accumulation? Perhaps, according to some, Earth's prime is past and the proper role of humanity is to hasten the degradation of this planet's natural environment.

It is easy to see from where such pessimism comes and how such defeatism has evolved. Our environment is degrading, impacted by human-caused pressures of overpopulation, rapid industrialization, and habitat destruction. One has to ask if the peaceful Eden of the theologian and the utopia of the philosopher ever actually existed? To investigate these questions, let's pick up the story after the appearance of the first life on Earth. In this chapter we will attempt to describe billions of years of history as a narrative, often claiming as fact scientific hypotheses in many areas of evolutionary biology, planetology, and history. While it may be that some of these events did not occur exactly as we describe, they are nonetheless reflective of many prevailing scientific theories.

First, Earth had an atmosphere in this era, but oxygen was a rarity. The comet- and volcano-supplied early terrestrial atmosphere may have been rich in nitrogen and carbon dioxide and there may have been ample hydrocarbons as in the modern atmosphere of Saturn's satellite Titan (Fig. 3.1), but free oxygen was either very rare or nonexistent.

Early terrestrial life not only survived without oxygen, these anaerobic (oxygen-hating) forms actually thrived in their environment. But life operates according to Darwinian evolution. Environments change, organisms can

L. Johnson et al., *Paradise Regained*, DOI 10.1007/978-0-387-79986-5_3,
© Praxis Publishing Ltd, 2010

Figure 3.1: A mosaic of Titan's surface, as photographed by the descending European Space Agency (ESA)/NASA Huygens Probe. (Courtesy of NASA.)

mutate, and the descendants of those best suited to an altered environment come out on top.

Perhaps it was 3.8 billion years ago when a crucial mutation occurred. Perhaps challenged by more successful forms, a microscopic anaerobe that had retreated to the upper layer of Earth's young ocean learned to gather energy from sunlight. In this process of photosynthesis, this plant-like organism converted solar energy and carbon dioxide into glucose, producing oxygen as a waste product. Over time, this new life form began to thrive.

This was perhaps the most critical of Earth's natural environmental catastrophes. The anaerobic forms must have felt that their Eden was slowly becoming a sewer or a toxic waste dump as oxygen became more and more prevalent in the atmosphere. Perhaps with a slight sense of desperation, they retreated to volcanic vents on the deep ocean floor, only to trouble the surface world during periodic, noisome El Nino events.

The surface layers of Earth's oceans now belonged to oxygen-loving microscopic, one-celled life forms. Competition and predation must have been fierce. As Earth slowly developed its oxygen atmosphere, mutation and natural selection gave rise to multicellular colony organisms similar to modern jellyfish.

Some of these colony organisms, perhaps to avoid being devoured by their stronger, nastier neighbors, developed specialized cells that in turn produced hard, protective shells. This was the great era of the trilobites and other early shellfish. One of the few present-day survivors of this period is the horseshoe crab.

Hardy plants were beginning to spread from the ocean shore inland. And insects that played the role of pollinators accompanied them. Vast forests covered the planet. The buried organic remains of this so-called Devonian Era formed over tens of millions of years and were converted by geological processes into coal and oil. We are consuming this wealth of our planet at a prodigious rate; it will be gone within a very few centuries. The geologically stored carbon released in a mere instant (from our planet's point of view) currently threatens global climate stability.

But evolution in the oceans continued at an ever more rapid pace. Somewhere at sea, a shellfish mutated to develop a crude interior skeleton. The first armored fish—the ancestors of modern sharks—had evolved. Nature's new trick—the backbone—soon caught on. The seas filled with vertebrates.

This takes us to 300 million years ago. Earth's plant-produced oxygen layer was fully operational. In the stratosphere, solar radiation energized chemical reactions that converted some oxygen molecules into ozone. With an ozone layer in place to filter harmful ultraviolet sunlight, the stage was finally set for vertebrates to emerge from the sea.

Everywhere, the strong preyed upon the weak. Civilized protections were absent in the natural state. In tidal pools along the shores of Earth's oceans, small fish took refuge from their fierce neighbors.

As the moon orbited Earth and Earth spun on its axis, lunar and solar tides periodically varied water levels in these fragile habitats. Somehow, a fish developed crude lungs so it could spend a portion of its existence above sea level, and therefore survive extreme tidal variations in water level. This very beneficial mutation was passed to this creature's offspring. Amphibians began to emerge from the seas. In the fullness of time, some of these creatures altered so they could live full time on land. These were the ancestors of reptiles, dinosaurs, feathered dinosaurs (birds), and mammals.

Around 200 million years ago, the warm forests of Earth belonged to the big guys. Huge plant-eating dinosaurs browsed on the vegetation. Smaller, smarter carnivores—the raptors, which would give rise to modern birds—silently stalked the huge ones.

Hiding in the undergrowth, doing their best to avoid their ferocious, gigantic neighbors were small, rat-like mammals that were the distant ancestors of elephants, tigers, and humans. They emerged briefly from their warrens to grab a dinosaur egg or two, and then scurried for cover.

Life was abundant in these times, and life was diverse. But at least from the viewpoint of the mammals, life was no picnic and Earth was no Eden. And so it might have remained, as Earth spun upon its axis and danced around its star. But change seems to be built into all natural systems. After more than 100 million years of stability, catastrophe rained from the skies.

Even without human intervention, bad things happen to our planet's biosphere. At intervals of tens of millions of years, occasional mass extinctions occur. During these events most terrestrial organisms perish and many terrestrial species disappear. When the dust finally settles, a remade biosphere emerges, one with new life forms and new ecological niches.

Some of these events are perhaps produced by the eruption of super-volcanoes. Compared with these, historical eruptions such as Krakatoa, Vesuvius, and Thera seem like mere sneezes. Volcanic dust from these blow-ups, suspended for decades in Earth's upper atmosphere, would block sunlight and result in unending winters.

Other mass extinctions in prehistory may have had cosmic causes. Certain types of exploding stars might have bathed our planet in a sea of gamma rays, fatally irradiating all but the hardiest specimens of terrestrial life.

About 65 million years ago, as the herbivores grazed under the watchful eyes of the giant raptors, as tiny mammals briefly emerged and immediately retreated for cover, another type of celestial catastrophe occurred. Perhaps some of the doomed animals glanced at the brightening sky, unaware of what the celestial spectacle signified.

From the sky, a chunk of ice or rock descended toward Earth's surface. This 10-km-wide fragment of asteroid or comet streaked across the sky and toward the ground at more than 10 km/sec. In the dim brains of the doomed dinosaurs, the aerial show must have seemed like a second Sun, dashing rapidly across the sky (Fig. 3.2). The explosion was like a million hydrogen bombs igniting in the same place and at the same time. A towering mushroom cloud that reached to the stratosphere replaced the enormous fireball. Shock waves raced across the plains and through the forests. Enormous firestorms devoured most vegetation. Towering tsunamis raced through the oceans, radiating outward from ground zero.

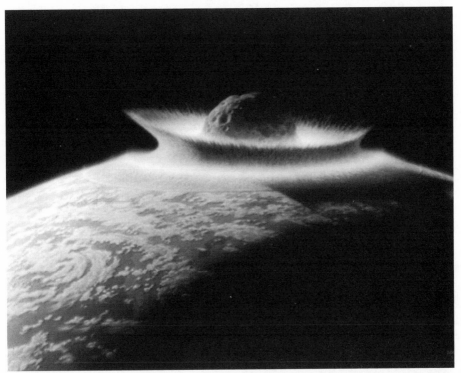

Figure 3.2: The great impact at the end of the Cretaceous Era would end the reign of the dinosaurs, but open the way for mammals including humans. Perhaps the event registered on some dinosaur brains before they were snuffed out. (Courtesy of NASA.)

Various seismic events such as volcanoes and earthquakes compounded the damage to living organisms, even those separated by geography from the direct effects of the blast. Higher life, at least, was erased from most of what would later be known as Earth's Western Hemisphere.

Earth was enshrouded in a vast halo of dust. For years or decades, temperatures plummeted globally as the dust layer reflected sunlight back to space. Most surviving vegetation perished with the onslaught of the cold, followed by most of the herbivores who fed upon it. Carnivores who had survived the impact may have initially had a field day, feeding upon the bodies of the deceased plant eaters. But they died as well since their food source could not be replenished.

Decades after the impact, skies began to clear to reveal a greatly altered world. Perhaps because they could hibernate, perhaps because small size had allowed them to find shelter, or perhaps because they were lucky, a handful of mammals emerged. They were accompanied by the ancestors of modern birds—small, feathered flying dinosaurs—that survived perhaps due to the thermal insulation of their feathers and their ability to fly from a degraded environment to a better one. Our distant ancestors may have dominated the landscape in the early Tertiary Era, as buried seeds began to sprout to re-germinate Earth's ruined forests. But life remained difficult.

As the eons flew by, the mammals mutated to fill ecological niches vacated by their vanished rivals. Creatures much like elephants, whales, deer, and apes ultimately evolved. As they competed for food and territory, evolutionary pressure picked up.

A few million years ago, an arboreal ape-like creature had evolved in central Africa. With opposable thumbs and a comparatively large brain, these creatures were well suited to life in the canopies of tall, forest trees.

But the environment changed again. As forests were replaced by prairie, some of these organisms descended from their perches to attempt life in the hostile surface environment. They faced great odds as they attempted to avoid the predations of the great cats and others. They could not know how momentous their descent from the trees was, but these humble beings were the ancestors of us all. In the rough-and-tumble environment of the African savannahs, competition would spur tool use and a rapid increase in brain size.

A few hundred thousand years ago, the first true humans evolved. No longer would they be victims of environmental change; the fate of the entire terrestrial environment would ultimately rest in their hands instead.

Further Reading

For a very readable survey of life's origin and development on our planet, see John Reader, *The Rise of Life* (New York: Knopf, 1986). The devastating impact that ended the reign of the dinosaurs is discussed by Carl Sagan in *Pale Blue Dot* (New York: Random House, 1994).

Paradise Lost?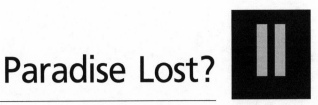

The Environmental Dilemma: Progress or Collapse?

4

O me! O life! Of the questions of these recurring,
Of the endless trains of the faithless, of cities filled with the foolish,
Of myself forever reproaching myself, (for who more foolish than I,
  and who more faithless?)

Walt Whitman, from "O Me! O Life!"

As we have seen, the distant past was difficult for the ancestors of man. But where have the development of civilization, the agricultural revolution, the art of metallurgy, and the scientific revolution left us? We seem to be suspended from an environmental cliff. We cannot hang on, but we are doomed if we let go!

Our global civilization may be the first to understand the dilemma of the environment. But we are not the first to be threatened by environmental collapse. Rather than endlessly reproaching ourselves about the foolishness that led to our predicament, it may be helpful to briefly review humanity's prehistoric and historic interaction with the environment.

Early in human prehistory, our environmental footprint was minimal. For about two million years, our ancestors lived off whatever nutritious local vegetation was in bloom, ate meat from whatever they could catch or steal from larger predators, and did their best to avoid the jaws of the bigger, fiercer animals. Without permanent habitation, it was not hard to escape a degraded environment. Our ancestors were effectively nomads; they would simply move elsewhere. But perhaps because they were slower and less well armed than the large cats, bears, or wolves, our human progenitors began to depend more on cleverness and stealth than on physical strength. As their brains developed, so did their tools.

With the aid of spears and arrows, they developed the skills to cooperatively hunt large game and to kill at a distance. The unknown genius that first tamed fire opened the way to the ultimate spread of humans from the tropical African home to temperate and arctic climes.

Perhaps their tribe was pursued by more powerful neighbors; perhaps it was population pressure or environmental change. Whatever the cause, a

L. Johnson et al., *Paradise Regained*, DOI 10.1007/978-0-387-79986-5_4,
© Praxis Publishing Ltd, 2010

small group of humans found itself hemmed in by the ocean, on what is now the southeast coast of Asia. More than 40,000 years ago we may speculate that one of them observed that fallen trees floated in water after a storm. Following this person's brilliant hunch, the tribe strung logs together to construct the first ocean-going rafts. Island hopping across the shallow sea, the people of this tribe ultimately arrived in what is now Australia.

Strange new animals inhabited a landscape covered with unfamiliar vegetation. After a few generations, these progenitors of the modern Aborigines learned by observation how effective fire was in improving fertility and clearing undergrowth. It may have been a tool to improve the hunting ground by thinning the forest. For the first time, humans were setting controlled fires to alter their environment. Ultimately, this would lead to what became "slash and burn" agriculture. It represents what may have been the first large-scale manipulation of the environment by humans.

In other parts of the world, as climate cooled and glaciers advanced, our human ancestors became well adapted to the Ice Age. From cave paintings in northern Europe, we have a fair idea about the methods used by hunting bands to bring down large game animals such as the now-extinct mammoth.

About 13,000 years ago, human bands migrated from Asia across a narrow land bridge connecting Siberia and Alaska. Modern humans had invaded the Western Hemisphere and would soon claim it for themselves. Interestingly, biodiversity in the New World declined a bit as certain large-animal species became extinct, possibly due to the hunting skills of the new settlers.

Then, the climate began to warm. The Bering Sea land bridge was submerged as melting glaciers raised ocean levels. In Europe, Ice Age people were faced with a major dilemma. As the ice retreated, so did the big-game herds that the paleo-Europeans depended on. Unless people were to follow the herds, radical alterations in lifestyle were required.

Not all near-humans were able to adapt to the warming climes. Our close cousins, the Neanderthals, vanished with the retreating glaciers. It may have been with timidity, fear, or trepidation that the once dominant (male) hunters approached the more submissive (female) gatherers about possible solutions. But about 10,000 years ago, in the Middle East and central Turkey, something brand new began to spread across the face of the Earth.

The Agricultural Revolution

This was the agricultural revolution, which began at the dawn of the Neolithic period—the New Stone Age. For the first time, in Jericho, Catal

Huyuk, and a host of other sites, humans were putting down permanent roots. We learned how to farm, and how to husband domestic animals. No longer were we at the mercy of the seasons, always pursuing the animal herds.

But there were other problems as populations of the early towns rose to the hundreds and thousands. One was pre-Neolithic nomadic bands, which might be tempted to rob our ancestors of Earth's bounty. So defensive walls were constructed around the growing towns. Organized warfare, which may have existed for millennia, became a major human endeavor.

There were other problems as well. In the beginning at least, early agriculturists were at the mercy of the elements. Too little rain would parch the crops; too much rain would drown them. So a mythology developed that connected humans to nature.

There were male sky gods such as Zeus, Jupiter, and Thor. They were responsible for the defense of the growing settled communities against the outside intruders. Other male gods—underground deities such as Poseidon and Neptune—were responsible for large-scale catastrophes such as earthquakes, volcanic eruptions, and tsunamis. Elaborate rituals were devised to keep these dangerous beings at bay.

But this was also the era of the Great Goddess. Female deities were in charge of the monthly and seasonal cycles. The greatest of these was Gaia, the goddess of Earth. Until about 1000 BC, when the Age of Stone that had yielded to the Bronze Age, was itself surpassed by the Iron Age, most civilized humans felt tied to the cycles of nature. Some of the great stone-circle observatories they used to keep track of seasonal cycles—notably Stonehenge in the United Kingdom—can still be visited today.

In this long era, people began to utilize the powers of nature in an unprecedented fashion. The early log rafts of the Paleolithic Australian migrants were rendered obsolete when some budding navigator observed the behavior of large water birds, perhaps swans, on a river such as the Nile. If such a creature desires to move against the river's current with the wind at its back, it simply fluffs its feathers and glides across the waters.

By 4000 BC, crude sailboats carried trade goods between communities located along the Egyptian Nile. Soon, some of these ventured into the Mediterranean. The Cycladic and Minoan civilizations further developed the sailing craft, with innovations including the keel, which allows sailboats to tack against the wind. Before the end of the Bronze Age, some of these craft ventured out into the Atlantic Ocean, to visit areas that are now England, Scotland, Wales, and Ireland. Before the Golden Age of Athens and the rise of Rome, crews of other vessels were exploring and trading along the coast of Africa. Sometime around 2000 BC, the death knell of this era resounded when someone on the Asian Steppes noticed that the horse, a delicious game animal, could be tamed for other functions.

The Age of Iron

The new partnership of human and horse opened the way for the ascendancy of the male, warlike sky gods. Carrying the banners of these deities and sophisticated weapons of bronze (and later of iron), chariot-equipped armies fanned out from their central Asian homelands to dominate the Indo-Pakistan subcontinent and what is now Greece. Myths about the hybrid human-horse centaurs reflect this turbulent era of prehistory.

City-states began, in certain parts of the world, to be incorporated into a new form of human organization—the centralized empire. Beginning in Africa and Asia with the rise of Egypt, China, and Babylon, the cult of the conquering warrior ultimately spread through Europe.

At least some of this centralization was a response to human civilization's interaction with the environment. Along the Nile River, human settlements grew in a narrow band of fertile land that was nurtured by the river and always threatened by the encroaching desert. As agriculture succeeded, the human life span increased and population expanded. Cities required rivers as both a fresh-water source (upstream) and a sewage dump (downstream).

This arrangement worked fine as long as there were few cities. But as human habitation expanded along the Nile, a centralized government was necessary to regulate water use for irrigation and waste disposal. The great Egyptian empire, which was incorporated more than 5000 years ago and served as a model for many future states, was very likely a response to environmental pressure.

As the ancient empires flourished, warred with each other, and were ultimately incorporated in the all-encompassing empire of Rome, people may well have noticed a change in human relationship with the world around them. No longer were people totally at the mercy of the elements. Humans were supreme, dominating the landscape as well as other life forms.

Sadly, Iron Age humans did not always use their new powers for humane purposes. An early environmental outrage occurred when Rome, after having finally defeated Carthage in the Punic Wars, sowed the soil around Carthage with salt in 146 BC to destroy the land's fertility.

There were setbacks to human ascendancy of the Earth, such as earthquakes, volcanoes, and possible meteorite impacts, but the devastation was local. With the rise of Christianity and Islam, the powers of humanity grew, as did the population.

One problem faced by the expanding cities was the diminishing local supplies of firewood. People required this resource for heat in winter and to cook food. But forests mature over periods of decades or centuries, so growing populations needed to travel farther and farther from home to obtain firewood.

During the Middle Ages, someone in England, Ireland, or Scotland realized that peat, an organic-rich soil, could be dried and used for fuel. Before 1600 AD, peat had been supplanted in cities such as London by a new fuel—coal.

Early Air Pollution Episodes

The new fuel had the advantage of freeing the growing population of London from a dependence on diminishing local forests. But this societal advance was a two-edged sword. Coal was generally burned in household hearths. Scrubbers and other pollutant-removal techniques were unheard of. Soon, an omnipresent dark cloud reduced sunlight levels. Exposed structures open to the elements began to feel the effects of corrosion. Worst of all, humans began to suffer from a host of respiratory ailments that had previously been rare or unknown.

As we now understand, atmospheric processes render air pollution a non-local phenomenon. Fallout from the sulfur-bearing effluent clouds produced in the cities traveled into the countryside, affecting humans, crops, and other organisms far downwind from the concentrations of coal-burning hearths and furnaces.

This type of air pollution, characterized by sulfur oxides and particulate matter, is fittingly called "London smog," after the city in which it first became evident. Modern centralized coal-burning power plants overcome much of this problem by employing scrubbing and filtering techniques to remove most of the sulfur compounds and particles from the effluent plume. But there are limits to the effectiveness of pollution control, no matter how sophisticated the technologies we employ.

The Modern World—Running Up Against Environmental Barriers

Four centuries after the large-scale introduction of fossil fuels (coal and later petroleum), our global civilization runs on these hydrocarbons. These remnants of ancient forests, modified by geological processes for millions of years, are being consumed by our global civilization at an ever-increasing rate and will be completely gone within a century or two.

In part because of this comparatively plentiful and inexpensive energy source, a larger fraction of humans live better than ever before. Our life spans are longer; a greater percentage of children (at least in the developed world) survive to adulthood than in any previous era.

But the laws of nature conspire to limit the longevity of our consumptive global civilization. To appreciate what is going on, we must make a short detour into that branch of classical physics called thermodynamics. Thermodynamics basically means "motion from heat." It essentially describes how to obtain useful work from an idealized heat engine.

Figure 4.1 is a schematic representation of such a device. Energy flows downhill from a hot reservoir to a cold reservoir. Think, for example, of your home furnace or an automobile propelled by an internal combustion engine. Fuel is burned at as high a temperature as possible (the hot reservoir), and waste products of this combustion are expelled to the much colder environment (the cold reservoir). Somewhere in this process, some of the energy produced by the combustion is diverted to heat your home or turn the auto's wheels.

The first law of thermodynamics is often stated in gambler's parlance as "you can't get something for nothing." This means that any heat engine functions in the real world; all useful energy supplied by such an engine comes from heat. This is reasonable; otherwise, a heat engine would be a perpetual-motion machine, magically supplying useful energy without fuel.

But the second law of thermodynamics is even more restrictive: it states, again in the language of the gambler, "you can't even break even." No heat engine will ever be 100 percent efficient. Some waste heat will always be expelled to the environment. From a mathematical point of view, efficiency increases as the hot-reservoir temperature increases and the cold-reservoir temperature decreases.

Many contemporary fossil-fuel power plants have efficiencies of around 50 percent. Since nuclear-fission plants often expel waste heat into enclosed water-filled cooling towers to avoid radiation release to the environment, they are generally slightly less efficient than fossil-fuel power plants that expel their waste heat to cooler natural water bodies.

To reach efficiencies as high as 70 percent, it is necessary to employ the technologies of magnetohydrodynamics (MHD). In experimental MHD plants, the high-temperature reservoir operates at about 1000 kelvin (K). At such high temperatures, the circulating fluid is an ionized gas or plasma rather than a gas such as superheated steam. A major obstacle to the wide-

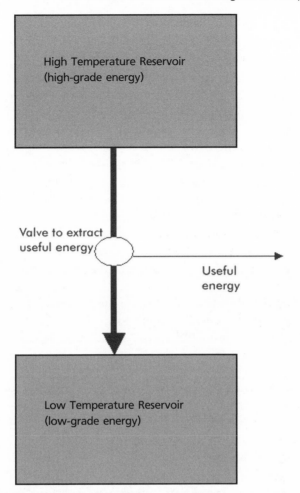

Figure 4.1. Representation of a heat engine, showing the extraction of useful energy as heat flows from a higher temperature to a lower temperature.

scale utilization of high-efficiency MHD technology is the very real possibility of injury or death if an accident caused the high-temperature plasma to escape to the environment.

The second law of thermodynamics is often called "entropy," which means that the disorder of the universe is always increasing. For the entire universe, this can be visualized by realizing that all the potential for (high-grade) energy production was concentrated in the interior fusion-fuel sources for infant stars in the universe's early eons. As the universe winds down in many billions of years, most of this energy will have been transferred to (low-grade) random motions of interstellar molecules and atoms—heat.

When applying the second law of thermodynamics to life, one realizes that considered in isolation from its environment, a life form is negentropic; that is, it learns more and remembers more as it matures. But this increasing order comes at the expense of the environment. Fortunately, Earth is not a closed system and it is bathed with enormous amounts of energy from the sun every day. It is from this sunlight that life derives the energy necessary to create and maintain itself.

Also, localized resource depletion is a recurring theme throughout the history of human civilization. If you consult the Bible or Homer's epics, you will learn that in the Bronze Age, Lebanon and Crete were famous for their old-growth forests. But 4000 years of high civilization has seriously degraded the natural environments of these and other ancient sites. Closer to home (in time if not in space) the paleo-American cliff dwellings in the American southwest were abandoned about 1000 years ago. Increasing population is thought to have placed too much of a strain on the limited water reserves in this arid environment for that culture to thrive and prosper.

In ancient times, Earth's human population was considerably lower than it is today. It was possible to think of the planet as an infinite sink for pollution. If your environment became degraded, it was readily possible to find another home elsewhere. But in today's heavily populated global society, we do not have the luxury of picking up roots, abandoning our cities, and migrating to fresher climes. Although our current environmental crisis was generations in the making, it came into clear focus in the 1970s with the Apollo photographs of Earth taken from deep space (see Chapter 12, Figure 12.1). For the first time, Earth was recognized as a fragile oasis of life suspended in a mostly sterile cosmos, a precious jewel rather than an infinite sink for civilization's waste products.

We are no longer completely at the mercy of the environment. The truth is that humans, as stewards of Earth, must develop an enlightened approach to the consideration of environmental issues if our civilization is to continue to thrive.

One approach is to broaden our definition of environment to include the solar vicinity of our planet. If energy and other resources can be obtained from space and some of the waste products of civilization (heat, at least) disposed of in the sterile environment above the atmosphere, the chance exists to greatly enrich the living standards of most people, without jeopardizing earth's biosphere. But accessible space itself is not infinite, as was demonstrated by the collision of two near-Earth artificial satellites in early 2009 and the resulting cloud of orbital debris.

Instead of viewing the immediate future with gloom, perhaps some optimism can be generated. If we use the tools at our disposal and rationally plan for the future, the human course may be bright indeed.

Further Reading

For a very readable account of human prehistory and history between 35,000 BC and 500 AD, see Jacquetta Hawkes, *The Atlas of Early Man* (New York: St. Martin's Press, 1976). Of the many books that discuss historical and modern pollution episodes, a good one is Laurent Hodges, *Environmental Pollution*, 2nd edition (New York: Holt, Rinehart and Winston, 1977).

Many sources discuss application of thermodynamics to energy and environment. A fairly up-to-date and beautifully illustrated text considering this topic is G.T. Miller, Jr., *Environmental Science*, 4th edition (Belmont CA: Wadsworth, 1993).

Exploding Population 5

Crowds of men and women attired in your usual costumes,
  how curious you are to me!
On the ferry-boats the hundreds and hundreds that cross,
  returning home, are more curious to me than you suppose,
And you that shall cross from shore to shore years hence are more
  to me, and more in my meditations, than you might suppose.

Walt Whitman, from "Crossing Brooklyn Ferry"

As Walt Whitman watched the multitudes commuting between Brooklyn and Manhattan in the days before the construction of the Brooklyn Bridge, he dreamt of the multitude that would make that crossing in the future. He could not have imagined the much larger population that has been born in recent years to challenge the future of Earth.

It is not uncommon for science authors to treat the population explosion as a form of plague, and to consider more billions of humans a threat. Here, we take the opposite tack: this is a challenge to be overcome. But if necessary changes to human lifestyle are made, healthy and productive lives are possible for the burgeoning human population. The opportunity exists to make these changes without desecrating our planet.

One thing to remember is that population levels have varied throughout history and prehistory. The number of humans existing at any time in our species' history has a lot to do with our interaction with the environment.

Human Population: The Prehistorical/Historical Record

No one can say that two million years ago, as proto-humans descended from their original treetop homes, there were too many people. In fact, the opposite may have been true. Because of the comparatively low level in human genetic variation, the original proto-human population may have been very low. Perhaps we are all descended from a few thousand individuals.

L. Johnson et al., *Paradise Regained*, DOI 10.1007/978-0-387-79986-5_5,
© Praxis Publishing Ltd, 2010

Many things could reduce the life span of an early human: death in childbirth due to nonexistent sanitation, infant mortality due to infectious diseases, culling of the young and elderly by hungry predators, and so on. The entire prehuman population in the early Paleolithic Era was centered in tropical Africa and probably numbered 100,000 or so.

When fire was tamed a bit later, people could migrate to cooler climes and increase their numbers somewhat. The staple diet now consisted of cooked meat, which may have slightly increased the human life span since most food parasites are killed by cooking. But there were other hazards to contend with; nontropic winters must have been fierce during periods of global glaciations.

By the late Paleolithic Era, the human population must have numbered something in the few millions. Even though people lived in small hunting groups of one hundred or so and the landscape was sparsely populated, humans spread around the world.

A significant increase in human population occurred with the onset of agriculture, around 10,000 years ago. Some of the Neolithic villages must have numbered in the thousands. Farming and animal husbandry meant that people living in settled communities enjoyed a more stable food supply than did their nomadic cousins. This must have resulted in a slight increase in average human longevity, but very few people reached the age of 50.

Metallurgy came into use with the Bronze Age, which extended from about 3000 BC to 1000 BC. At least some people now lived in true cities with populations of 10,000 or more. With such large populations and metal tools, sanitation became a major endeavor. Delivering fresh water to a population center and removing sewage became the responsibility of a new class of civil engineers. Organized medical procedures also date from this period, and this also contributed to increased longevity. Famines in this period were alleviated as the new god-kings learned to store grain in good years and distribute it to their subjects when crops failed. There were as yet no global population counts, but tens of millions may have inhabited our world.

With the arrival of the Iron Age at about 1000 BC and the rise of the empires, population levels in major cities may have exceeded 100,000. Even with advances in hydrology, sanitation, and food distribution, a new limiting factor arose to reduce human population levels. As yet, there was no understanding of microscopic pathogens. Various plagues afflicted the human population, which were more effective than warfare in limiting human numbers.

All through the Middle Ages, the human population slowly increased, except when incidents such as bubonic plague—the so-called Black Death—caused human numbers to crash. But the total global population never exceeded a few hundred million.

It should not be supposed, even for a moment, that medieval Europeans were innocent victims of the bubonic plague. In a frenzy of irrational brutality, the church-led Inquisition had slaughtered multitudes of women as witches. Many cats were also thrown to the flames, suspected of being witches' familiars. If these blameless creatures had survived, they might have killed more of the rats that hosted the parasite responsible for the deadly disease.

But the fundamentalist fanatics did not have the last word. Around 1500 AD, rationality reentered the human arena with the Renaissance. As classical learning was rediscovered, the scientific method was perfected. For the first time, a philosophical formalism existed whereby hypotheses about nature would be tested by observation and experiment.

Although the broadening of humanity's worldview by the outward-peering telescope pointed the way to the new science of physics and its attendant, world-creating technologies, it can be argued that the biological revolution spurred by the microscope was even more profound. As early biologists used this new tool to observe the behavior of organisms too small to be seen by the unaided eye, a new theory of disease began to take hold. As this germ theory matured, it became impossible to maintain the ancient concept that disease was caused by improperly balanced humors within the afflicted organism. Microscopic entities—first bacteria and later viruses—were pinpointed as the causes of disease.

By the year 1900, the human life span had taken another leap forward. Sterile conditions in the operating room did wonders to extend the life of typical humans in developed countries. No longer would a young woman expect to die in childbirth; no longer would a family have six children so that two might survive to maturity. Around this time, the human population approached or exceeded the one billion mark.

The twentieth and twenty-first centuries have witnessed an explosive increase in human population. In early 2009 there were more than six billion people on Earth. The population may peak at around ten billion before 2050. This is a remarkable development. Even in light of destructive wars (such as World War II, which resulted in about 60 million deaths), newly evolved plagues such as AIDS (which has killed tens of millions), and many wars entailing episodes of "ethnic cleansing," the population continues to increase.

Population Increase: Its Causes and Ramifications

It has been pointed out that human population levels have followed a J-shaped curve—nearly flat for thousands or millions of years, followed by a

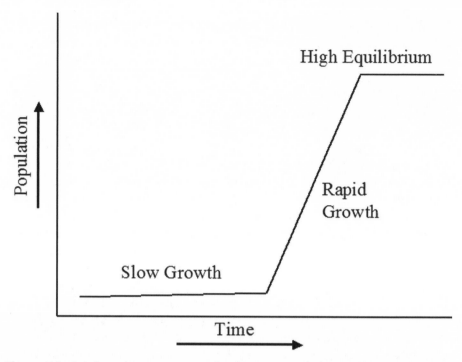

Figure 5.1. A J-shaped curve representing human population variation with time.

very rapid increase, with a new, higher equilibrium level achieved at the end of the rapid increase. An example of such a J-shaped curve is shown as Figure 5.1. Will population remain constant at around ten billion, continue to increase, or crash to a much lower value?

One reason for the dramatic rise in human population is the success of modern medical science. With our current understanding of the evolution and behavior of human pathogens, drugs have been developed that have dramatically increased the human life span. Unhealthy behaviors have been altered (at least in the developed world), sterile medical procedures are the global norm, and knowledge regarding nutrition and nutritional supplements has become widespread. Another success of civilization has been elimination or reduction of the larger predators who once preyed upon us.

Population dynamics have a number of interesting aspects. One might think, for instance, that an average married couple must have two offspring to maintain a constant population. However, a certain percentage of every generation will not marry, and some children will die before they are physically mature enough to reproduce, even in developed countries. For these and other reasons, married couples must have an average of about 2.3 offspring to maintain a constant population.

In the developed world, where there are many employment opportunities for women outside of the home, typical couples often have less children than the number required for a stable population level. If immigration were to be efficiently reduced or eliminated, populations in these regions might actually decrease over time.

In many parts of the developing world, population is still increasing rapidly. One reason for this is the sociological inertia of many traditional cultures.

Before the advent of modern medicine, a typical married couple might have six or more children to ensure that two might survive to become adults. Modern medicine has drastically reduced infant and child mortality, but adherence to traditional family values in these regions still results in large families, especially among the rural poor. Very draconian measures, such as those practiced by the Chinese authorities, are implemented in many of the less developed countries to reduce population growth.

Must Population Ultimately Crash?

A peak global population of about 10 billion people around midcentury is a likelihood. But what happens next?

One person who contended with ultimate limits to population growth was Thomas Malthus, a late 18th-century British economist. According to Malthus, agricultural output generally increases arithmetically. For example, a farmer might have 10 acres under cultivation in 2010. He might add 1 acre per year so that he has 15 acres under cultivation in 2015.

But population increase in Malthus's era was geometrical. To gain an understanding of geometrical increase, consider the fable of two men competing in a chess game. The winner requested that his prize should consist of one grain of rice on the first square of the chessboard, doubling the number of grains on each square thereafter: two on the second square, four on the third square, and so on. While it might seem that the losing player had gotten off cheaply, the progression results in far more than the world's annual rice production well before the loser gets to the final, sixty-fourth square of the chess board.

Even though Malthus used these arguments to predict that agriculture would ultimately fail to supply ample foodstuff and the population would decrease due to famine, such a dire future is not inevitable. Population growth does seem to be slowing as couples in developed countries have fewer children. Agricultural science has thus far proven effective in developing more efficient crop strains.

But there are more subtle forces to contend with. For instance, the expanding human population results in habitat destruction for other large mammals. As various mammal species face population reductions, the parasites infesting these organisms can mutate to change hosts. Since a numerous host is the human species, medical science faces a continual challenge that will only get worse as biodiversity decreases.

Space Technology and Human Population Increase

Since the subject of this book is the application of solar-system resources to maintain and improve Earth's environment, it is worth considering what space technology can do to alleviate the population explosion. Bur first we discuss what it cannot do. Wholesale migration of excess humans to cosmic sites will almost certainly not alleviate the population crunch.

Assume for the sake of argument that we eventually learn how to construct deep-space habitats for large numbers of humans using resources found on the moon or nearby asteroids. (This concept, after being proposed by Princeton University physics professor Gerard K. O'Neill in the 1970s, was then developed theoretically be numerous researchers.) If population grows by 100 million people per year, about 300,000 people would need to depart Earth and travel into space each day.

At present, if you wish to spend a week as a tourist aboard the International Space Station, the bill from the Russian launch services will amount to the equivalent of about $20 million. Let's say that breakthroughs and technological improvements eventually reduce this to $1 million, including baggage, life support, and other essentials to support the space immigrants and that the cost of producing the space habitats is reduced to zero, probably by application of a robotic labor force. The global space immigration cost per day would amount to about $300 billion—a somewhat daunting figure. The entire global economy—and more—would be devoted to the space immigration effort.

Furthermore, such an approach might fail to have the desired effect. The large-scale European emigration to the United States in the 19th and early 20th centuries did not substantially reduce the populations of European countries. New babies were rapidly produced to fill the gaps left by the emigrants.

But space-age technology, such as closed-loop life support, recycling, efficient housing concepts, and intensive agriculture, can also be applied on Earth. In the late 1960s, architect Paolo Soleri introduced the concept of arcology. An "arcopolis" is an Earth-bound version of a space habitat or

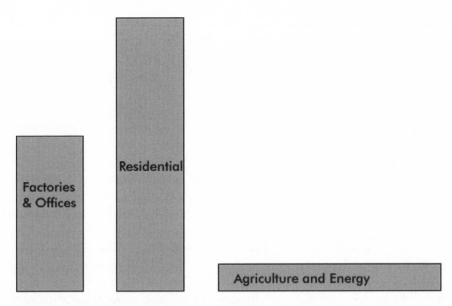

Figure 5.2. The relative benefits of an arcopolis.

interstellar ark, used to house vast numbers of people in a compact city state. Such a complex is represented schematically in Figure 5.2.

Let's consider the effectiveness of such a compact, highly urbanized lifestyle on overpopulated planet Earth. Assume first that each of the 10 billion people on the planet in 2040 requires about 300 square feet (33 square meters) of comfortable living space. If everybody lived in a one-story house, the human-habitation "footprint" would have an area of about 3 trillion square feet, or a million square miles, about 2% of Earth's land surface area. When we factor in the amount of land space that must be devoted to agriculture, energy-production, transportation, and so on, there is not much room left for rainforests and parks.

But consider instead that people choose to live in high-rise skyscrapers. Imagine a residential structure 1500 feet (500 meters) tall, a bit shorter than the world's tallest building. There are 150 stories in this hypothetical building. The building's footprint is 300 × 60 feet. Each story has enough floor space to comfortably accommodate sixty people, so the entire structure can house 9000 people.

From the NASA-funded follow-up work on O'Neill's space habitat ideas, the closed-ecological-system agricultural space required in space to feed one person is about 600 square feet (or about 60 square meters). Since Earth-based agriculture must contend with the day-night cycle and weather and seasonal changes, we will assume that 2000 square feet are required to feed each person. As in space, the assumption is made that water is efficiently recycled.

Our 9000-person arcopolis therefore requires a footprint of 18,000 square feet for living space and 18 million square feet (less than 1 square mile) for hydroponic agriculture. In the unlikely event that an entire global population lived in such structures, the total agricultural and residential space required to house and feed a population of 10 billion people would approximate 1 million square miles, which is about 2 percent of Earth's land surface area.

Since an arcopolis would be compact, the energy-inefficient automobile would not be widely used for commuting. Energy-efficient, high-speed rail networks could be used to connect urban centers, as is already the case in Western Europe.

Efficient application of available high technology, closed-environment organic farming techniques, recycling, and other technologies has the potential to allow comfortable living for a global civilization of 10 billion people. There will be plenty of room for parks, museums, theaters, sports stadiums, schools, hospitals, universities, and wilderness.

The reader might be happy to learn that Soleri's concepts represent an ongoing project rather than a dream for the far future. At the age of 88, Soleri works with the nonprofit Cosanti Foundation, which is developing a prototype ecologically sustainable community in the Arizona desert near Cordes Junction, about 65 miles north of Phoenix. As of June 2008, about fifty people live in this prototype "Arcosanti," which will house more than one thousand upon completion. Workshops, seminars, and conferences are conducted at the site, which is ideally suited to ultimately export wind or solar energy.

Located on marginal land for agriculture or conventional human development, Arcosanti is partially underground. Equipped with solar greenhouses on 25 acres of a 4000-acre preserve, it may serve as an urban laboratory for future experiments in high-density urban living.

Because construction and operation of this desert prototype arcopolis is not funded by government grants, the Cosanti Foundation has developed an innovative fund-raising strategy. About 65 miles south of Arcosanti, in Scottsdale, Arizona, tourists can visit Cosanti, which has been designated an Arizona historic site. Here, they can support the project (which may ultimately require $200 million) by purchasing bronze or ceramic Soleri wind chimes, which can easily be mounted in city or suburban backyards or courtyards and in rural sites.

The problems of implementing an arcopolis and related technologies over the next century are daunting but not insurmountable. The choice is collectively ours, but if we choose and plan wisely, the long-term future can be bright.

Further Reading

Many environmental references discuss population-related issues. Two of them are G. Tyler Miller, Jr., *Environmental Science*, 4th edition (Belmont, CA: Wadsworth, 1992) and Bernard J. Nebel and Richard T. Wright, *Environmental Science*, 4th edition (Englewood Cliffs, NJ: Prentice Hall, 1993). The first of these books introduces the J-shaped curve and presents the fable of the chess players.

Space habitat reference include Gerard K. O'Neill, *The High Frontier* (New York: Morrow, 1977) and R.D. Johnson and C. Holbrow, editors, *Space Settlements: a Design Study*, NASA SP-413 (Washington, DC: NASA, 1977).

Paolo Soleri first published his Arcology concepts in P. Soleri, *The City in the Image of Man* (Cambridge, MA: MIT Press, 1969). Updated material on this topic can be found at the following web site: www. arcology.com. To learn more about the ongoing work of the Cosanti Foundation, you can visit them online at www.cosanti.com.

gamma ra

radio continuum (2.5 GHz

x-ra

optica

near infrare

mid-infrare

infrare

molecular hydrogen

radio continuum (408 MHz)

atomic hydroge

Climate Change 6

Some say the world will end in fire,
Some say in ice,
From what I've tasted of desire
I hold with those who favor fire.

Robert Frost, from "Fire and Ice"

Here are some recent headlines from articles about global warming:

Arctic Melting Fast; May Swamp U.S. Coasts by 2099
National Geographic News, November 9, 2004

2005 is Warmest Year on Record for Northern Hemisphere, Scientists Say
USA Today, December 16, 2005

Study: Earth "Likely" Hottest in 2,000 years
The Associated Press, June 22, 2006

Unprecedented Warming Drives Dramatic Ecosystem Shifts in North Atlantic
Science Daily, November 7, 2008

These articles are alarming. They warn that Earth is getting warmer and changes in the global climate are inevitable. The evidence of a global change in climate is mounting and the consensus is that humans are responsible through our profligate emission of so-called greenhouse gases. The authors of this book are not climatologists and are therefore not qualified to assert whether or not humanity is responsible for climate change. But as scientists, we can say that the body of evidence being put forth by climatologists asserting climate change seems credible. The theory that the change is being caused by human activity is almost as compelling and merits serious consideration and concern. Before we go further, we should describe what we mean by climate change and review some recent history.

L. Johnson et al., *Paradise Regained*, DOI 10.1007/978-0-387-79986-5_6,
© Praxis Publishing Ltd, 2010

What Is Climate? Why Is Climate Stability Important?

According to Wikipedia, climate "encompasses the temperatures, humidity, rainfall, atmospheric particle count, and numerous other meteorological factors in a given region over long periods of time, as opposed to the term weather, which refers to current activity." So a rainy day in London is an example of weather while a propensity for rainy days in London might describe an element of its climate. Obviously, different geographic regions have different climates. The climate of Saudi Arabia is certainly different from that of Greenland and different still from that in Ecuador.

Climate is a driving force in determining where people live, their occupations, and often the relative affluence of those who live in any given location. Obviously, farmers in the American Midwest have a climate more favorable for agriculture than would be found in the African Sahara. Less obvious would be the connection between the rise and fall of entire civilizations wrought by changes in the climate. A report in the November 6, 2008, issue of *Science* discusses how the strength and frequency of Asian monsoons correlates with the rise and fall of the Chinese Tang, Yuan, and Ming dynasties within the last 1800 years. It seems that during strong monsoon periods, rain-dependent rice crops flourished and food was plentiful. When the monsoons became less frequent for many years (a climate shift), a correlation was observed between food scarcity and the fall of dynasties.[1]

According to Professor Gerald Haug of the University of Potsdam, Germany, and his colleagues, "Climate change is to blame for one of the most catastrophic collapses in human history." The collapse he was referring to was that of the Mayan civilization in Mesoamerica that ruled much of that region through about 800 AD. Then, after being afflicted with a succession of lengthy droughts, their empire with a population of 15 million people collapsed. The droughts, combined with deforestation and the resulting soil erosion, ultimately appears to have devastated the Mayan civilization.[2]

Will the climate change we are now experiencing change the political, economic, and military balance of power in our modern world?

Changing Climate Is Not New

Based on the evidence left behind, we know that Earth has gone through at least three or four ice ages with advancing then retreating glaciers. From valleys carved through mountain ranges to rock walls deeply scarred by moving mountains of ice, the unmistakable signs of a much colder Earth with more snow and ice are visible today. In addition, analysis of ice cores containing tiny bubbles of ancient atmospheres reveal signature characteristics of multiple cold, glacial periods followed by warmer and more temperate ones.

Earth's climate is a very complex system with many variables, which makes it very difficult to model. It is also difficult to determine the specific effects of changes in many input parameters with any certainty. There are some variables, however, that are so significant that they can easily dwarf all other factors and produce dramatic changes in the global climate. Examples of these significant factors include changes in the sun's intensity, volcanic eruptions, and changes in Earth's orbit.

Solar Variations

The sun may appear to be constant, but it is far from unchanging. Within its volume it could contain over 1 million Earths. The outer surface of the sun is about 6000 K (10,000°F), and, by nuclear fusion in its core, it converts approximately 5 million tons of matter into energy *every second*. This energy travels through space at the speed of light and a fraction of it impacts Earth, providing the heat and light that sustain our planet. There are sometimes explosions in the sun's atmosphere, called solar flares, that release enormous amounts of energy into space, often impinging upon Earth. The frequency with which these flares occur varies with the solar cycle. During the peak of the cycle, several flares may erupt in a single day. During solar minimum, there might be less than one per week. Despite these vagaries, Earth, on average, receives about 1.4 kilowatts of energy per square meter. If measured from space and integrated over all wavelengths, the energy striking the atmosphere of Earth varies by less than one tenth of 1 percent over an 11-year solar cycle. This continuous and mostly unchanging solar energy input into Earth's biosphere is called the solar constant.

As common sense might indicate, if the energy reaching Earth from the sun increases, the average global temperature will increase. If the energy reaching us decreases, the average global temperature will decrease. Within fairly recent recorded history there was a dramatic decrease in Earth's temperature that is attributed to a decrease in the solar constant. Between the 1400s and 1700s, solar activity appeared to be minimal, and many infer that the energy reaching Earth decreased. There are widespread accounts of the "Little Ice Age" that resulted.

European history records that during this period the Thames, Bosphorus, and other rivers froze, as did New York harbor. Northern Europe and North America experienced much colder summers with commensurately shortened growing seasons. All over the world there were reports of glaciers expanding and record snowfalls.

Some historians surmise that the much colder climate resulted in the demise of the Norse settlements in Greenland, paving the way for other Europeans to rediscover North America and make it their own.[3] One has to wonder how the history of the last 700 years might have been different if the sun had not gone relatively inactive and these settlements had prospered and grown. There might not have been a need for Columbus and his Spanish-funded voyages across the Atlantic.

Volcanoes

Volcanoes emit tons of ash and aerosols during eruptions. Large eruptions spew forth larger amounts of both, changing not only the local environmental conditions, but also climate worldwide. These dust and aerosols reflect an additional fraction of the sunlight falling upon Earth back into space, resulting in less energy actually striking the planet. One of the best-documented examples of this occurred in 1815 with the eruption of the Tambora Volcano in Indonesia. Some historical accounts record that the years following the volcano's eruption were up to 5°F cooler than normal, producing at least one year "without a summer." Needless to say, it is difficult to provide the population with enough to eat if a growing season is lost.

Orbital Variations

Over very long periods of time, small variations in Earth's orbit as it goes around the sun result in Earth being slightly further away from the sun and therefore receiving less energy from it. During those periods, the summers would not be as warm and the winter snows would not completely melt away. As anyone who has been outside on a sunny day following a snowstorm can tell you, it is blinding. The sunlight reflects from the white snow, depositing little of its energy within it. With more of the warming sunlight reflected, Earth grows cooler still, resulting in yet more snow accumulation. This "positive-feedback" cycle repeated over many years results in an ice age. When the orbital variations change still more, and if the amount of sunlight increases, then the result will be glacial melting and warmer global temperatures.

Climate Is Changing—Rapidly

Since 1850, and as of this writing, 2005 was one of the two warmest years on record, followed by 1998, 2002, 2003, and 2004. The Arctic Ocean, once covered with massive ice sheets year round, is melting at an alarming rate. According to satellite observations and measurements from shipping and aircraft records, the Arctic ice is well below its average level and is dropping fast.[4] Observations from the other side of the world show that the Antarctic ice sheet is gaining mass and getting larger.

Areas affected by sustained drought are growing as rainfall patterns change (see also Chapter 12). Snowfall in European and American mountain ranges is in decline. The months that bring rain to the American Southwest are shifting from October through April to October through March, resulting in a loss of one month's rainfall and increasing the length and severity of that region's fire season. These are but a few examples of how Earth's water cycle is changing before our eyes.

The onset of the greening of spring, when plants begin to sprout new growth, comes earlier almost every year. Farmers are adapting to the change and planting their crops weeks earlier than in the past. The scientific and anecdotal evidence for climate change mounts.

Are People to Blame?

There is evidence that human activity is responsible for the current period of changing climate. The prevailing theory is that our emission of greenhouse gases such as carbon dioxide is resulting in more heat being trapped by the atmosphere, thus increasing average global temperatures. According to the Intergovernmental Panel on Climate Change (IPCC) in 2007, the atmospheric concentration of carbon dioxide in 2005 was 379 parts per million compared to the preindustrial levels of 280 parts per million. Numerous studies indicate that this increase correlates with a rise in measured temperatures rather convincingly.

If this is a correlation, meaning that the two variables of temperature and atmospheric carbon dioxide levels just by happenstance change at the same rates and at the same times, then we have nothing to worry about. But if there is a causal connection, meaning that one (the level of carbon dioxide in the atmosphere) results in the other (the atmospheric temperature), then we humans had better take notice, as we are entering a period in which carbon dioxide in the atmosphere is increasing at an unprecedented rate. What, then, will be the ultimate impact on global temperatures? How will the climate change in response? The answer is that we do not know, but scientists have models that make some dire predictions.

As water warms, it expands and takes up more volume. Since Earth is a water world, with approximately 70 percent of Earth's surface covered in water, even a modest increase in sea temperature will result in the sea level rising. If you add the water from melting Arctic glaciers, then it will rise even more. The consequences of a dramatic rise in sea level will potentially include the destruction of coastal wetlands and barrier islands (making our coasts more susceptible to damage from tropical storms), and a greater risk of flooding in coastal communities. Some islands may become uninhabitable, resulting in the displacement of the people living on them.

If the climate warms, diseases once restricted to the tropics will spread to more temperate climes, infecting people previously not threatened by them. Diseases such as malaria and dengue fever are among those now observed to be spreading beyond their normal habitats. In addition, warmer temperatures can cause higher incidences of food-borne illnesses; the warmer temperature causes food to spoil more rapidly and some bacteria to reproduce more vigorously.

Studies showing the impact of increased temperatures on food crops yield similar bad news. A widely reported study by Lawrence Livermore National Laboratory and Stanford University estimates that the annual yield of wheat, rice, corn, soybeans, barley, and sorghum will decrease by 3 to 5 percent for every one degree of temperature increase.[5] Given that these crops are a significant source of food for much of the world, one can only surmise what might result, physically and politically, if these estimates are accurate.

Other species on planet Earth also will be affected by the increase in temperature. For example, when carbon dioxide mixes with water, it forms carbonic acid. As we pump more and more carbon dioxide into the atmosphere, more and more of it will interact with Earth's ocean water, forming carbonic acid as a result. The increased acidity of seawater damages ocean species with calcium carbonate shells as well as the coral reefs. Some scientists predict that up to 97 percent of the world's coral reefs could be destroyed if the global temperature rises just 3.6°F.

Some impacts of a warming Earth are counterintuitive. Take, for example, the case of the extreme Northern Hemisphere getting warmer. One would expect that warm-water fish of the middle Atlantic would be migrating northward as their ecosystem expands. But the opposite appears to be happening: fish normally found only in the cold waters of the north Atlantic are now being found further south than ever before. Why? Because melting Arctic ice sheets and glaciers are releasing cold water into the Atlantic, which is then carried southward by the ocean's currents. This colder water expands the ecosystem of the northern Atlantic fish, allowing them to venture further south than normal.

A review of the literature and the popular press will return more estimates, scientific or otherwise, postulating the negative (and some positive) effects of climate change. Unfortunately, there is a lot of hype surrounding the subject, with advocates of certain political views seizing on the issue of climate change to advance their agendas and potentially exaggerating the impact of climate change to suit their own ends. The converse is also happening. Those who do not want to make changes in the way they live or do business will take advantage of any scientific study questioning the reality of any aspect of climate change and attempt to use it to say that the threat is not real and that we should continue business as usual.

We take a more cautious view. Human beings can and do affect the environment and, potentially, the climate. If we can make changes in how we live and conduct business so that we are better stewards of Earth, then we should seriously consider doing so. Later chapters describe what space development can contribute toward minimizing the negative environmental impact of our advanced technological society. We cannot afford to ignore the risks, nor can we afford to ignore a path that might eliminate those risks.

Further Reading

To learn more about the effects on climate from the variations in Earth's orbit around the sun, see Milutin Milankovitch's seminal 1930 paper on the topic, "Mathematical Climatology and the Astronomical Theory of Climate Change." More information about Milankovitch is available in J.D. Macdougall's book, *Frozen Earth: The Once and Future Story of Ice Ages* (Berkeley, CA: University of California Press, 2004).

A comprehensive listing of the evidence supporting the reality of contemporary climate change can be found in K.E. Trenberth, P.D. Jones, P. Ambenje, et al., "Observations: Surface and Atmospheric Climate Change," in S. Solomon, D. Qin, M. Manning, et al., eds. *Climate Change 2007: The Physical Science Basis. Contribution of Working Group I to the Fourth Assessment Report of the Intergovernmental Panel on Climate Change* (New York: Cambridge University Press, 2007).

References

[1] "Climate Change: Chinese Cave Speaks of a Fickle Sun Bringing Down Ancient Dynasties," Richard A. Kerr, *Science*, 7 November 2008, 322: 837–838.

[2] "Climate and the Collapse of Maya Civilization," Gerald H. Haug, Detlef Günther, Larry C. Peterson, Daniel M. Sigman, Konrad A. Hughen, and Beat Aeschlimann, *Science*, 14 March 2003, 299: 1731–1735.

[3] "Interdisciplinary Investigations of the End of the Norse Western Settlement in Greenland," L.K. Barlow, J.P. Sadler, A.E.J. Ogilvie, P.C. Buckland, T. Amorosi, J.H. Ingimundarson, P. Skidmore, A.J. Dugmore and T.H. McGovern, *The Holocene*, 1997, 7: 489–499.

[4] "Satellite Evidence for an Arctic Sea Ice Cover in Transformation," Ola M. Johannessen, Elena V. Shalina, Martin W. Miles, *Science*, 3 December 1999, 286 (5446) 1937–1939.

[5] "Global Scale Climate-Crop Yield Relationships and the Impacts of Recent Warming," David B. Lobell and Christopher B. Field, 2007, *Environ. Res. Lett.* 2, 014002.

Vanishing Life 7

Roots and leaves themselves alone are these,
Scents brought to men and women from the wild woods
  and pond side,
Breast-sorrel and pinks of love, fingers that wind
  around tighter than vines,
Gushes from the throats of birds hid in the foliage
  Of trees as the sun is risen.

Walt Whitman, from "Roots and Leaves Themselves Alone"

When people think of the issue of decreasing biodiversity, they generally consider pets, domestic animals, majestic beasts, and useful plants. Although many beasts and plants are threatened (pets and domestic animals are not threatened), the problem runs far deeper.

There is an interconnected web of life and it is very difficult to cull one species without affecting others in the ecosystem. To further complicate things, there is not one ecosystem, but many. If we damage one, no one knows what the consequences may be for others.

But all is far from hopeless in our attempts to lessen humanity's harmful effects on the ecosystem. Coincidentally, our understanding of the ecosystem is growing as rapidly as our ability to alter it. More and more people view humanity as part of life's fabric, not as something outside of nature. This understanding may lead to the enlightened knowledge required to save many species great and small.

The Origins of Biodiversity

We know that there are many different species. But has this always been true? Will this always be true? Sadly, there is no real answer.

To investigate the origins of biodiversity, it is necessary to consider the origins of biological life forms. Scientists generally assume that life will originate naturally from nonliving systems given the proper conditions,

L. Johnson et al., *Paradise Regained*, DOI 10.1007/978-0-387-79986-5_7,
© Praxis Publishing Ltd, 2010

such as appropriate temperature, moisture, and nutrient mix. But as long as we have only one example of a living world, we may never know if this, or the rival hypotheses that life's origin is an exceedingly improbable random event or the result of divine intervention, is correct.

Even if life is confirmed on Mars, Europa, or some other solar-system world, the debate may not be resolved. It is not impossible that life could arise at only one location within a given solar system and be transferred by cosmic or geological events—meteorite impacts or volcanoes—to other planetary surfaces.

But what we do know is that fossils of terrestrial life exist, thus indicating that life originated on our planet almost four billion years ago. Within a few hundred million years of this origin, early life had radiated to form the progenitors of the first animals, plants, fungi, protozoa, and bacteria.

Long before the rise of mammals, long before the first human hefted the first spear, Earth teemed with myriad life forms. Today, human activities threaten this biodiversity. But geological and cosmic events have also threatened life in our planet's long history.

Natural Mass Extinctions

It is fashionable in some circles to believe that pre-human life existed in an idyllic state. But even in the most peaceful natural state, there is fierce competition among species and members of the same species.

All life is programmed to do its best to pass on its genes to future generations. The organisms that succeed are those best adapted to their environment. Failing organisms may vanish or become extinct. Random mutations will alter the genetic structures of the survivors so that vacant ecological niches in the biosphere will be filled.

Nature seems peaceful because the evolutionary process works in slow motion. Life spans of individuals are measured in years, decades, or centuries. But many species survive for millions of years.

Sometimes, however, the rate of change is accelerated. At intervals of tens of millions of years, some action reshuffles the genetic deck. Periodically, events occur that rapidly transform the terrestrial environment on a global scale. These are the so-called mass extinctions. We know that impacting celestial objects contribute to at least some of them. About 65 million years ago, an asteroid or comet about 10 km across slammed into what is now the Yucatan, in Mexico. Large dinosaurs and many other organisms vanished in the aftermath of this event, to be replaced by mammals and the feathered dinosaurs (more commonly called birds).

Since the sun orbits the center of the Milky Way galaxy, taking about 250 million years to complete one revolution, our solar system sometimes approaches star-forming regions. Exploding stars (also called supernovae) within these clouds may then bathe our world in gamma rays and x-rays, irradiating and extinguishing many life forms.

Natural terrestrial activities also cause some mass extinctions. Supervolcanoes many times larger than Krakotoa or Vesuvius have the same effect as celestial impactors; they enshroud the entire planet in a long-lasting layer of high-altitude dust that blocks sunlight and causes a precipitous drop in surface temperature.

When a mass extinction occurs, as many as 95 percent of terrestrial species may become extinct. It may be 100,000 years or so after the event that the surviving organisms finally rebuild a healthy, but greatly modified global ecosystem.

Our planet seems to be undergoing a mass extinction event. But this one is far different from those in the fossil record. We cannot blame an impacting celestial visitor. No supernova has occurred within our galactic vicinity, at least for many millions of years, and we cannot pin the blame on natural terrestrial events such as super volcanoes. Only one terrestrial species seems to be responsible for the extinctions. It is currently, without much foresight or planning, eliminating entire ecosystems at a prodigious rate. The guilty party is none other than *Homo sapiens*.

Natural Ecosystems

To better understand humanity's destructive role and to alter it, it is necessary to gain some understanding of the ecosystem. From deep space, Earth appears to be an integrated living organism. But on closer inspection, it has many separate components. Naturalists have attempted to categorize the various ecosystems that make up our living Earth. Ecosystems are those distinct environments, with the organisms adapted to surviving there.

In his 1984 epic book *The Living Planet*, David Attenborough listed and described the major natural distinct but interconnected terrestrial ecosystems. These include the oceans, seashore, fresh water bodies, volcanic calderas, polar ice sheets and tundra, forests, jungles, grasslands, deserts, and the sky. Also, because civilization has been widespread on planet Earth for 5000 years or more, we must include the semiartificial ecosystem including those organisms that have adapted to live near human concentrations.

Since the publication of Attenborough's book, scientific ecology has advanced. It is now understood that there exists another ecosystem, possibly more significant and elaborate than the ones listed above: the subterranean world of Earth's crust.

The web of life within and between the ecosystems is of varying strength. Humans, through over-fishing, can damage the deep-ocean ecosystem and influence the shore as well. But even if we are foolish enough to attempt such a deed, we cannot destroy all life on this planet. Even if we managed to wipe out surface, ocean, lake, and aerial life, subterranean forms would ultimately recolonize the planet's surface. But this recolonization might take tens of millions of years.

The Human-Directed Ecosystem

At least on a small scale, humans have been modifying natural ecosystems for millennia. At the dawn of the Neolithic, around 10,000 years ago, our ancestors domesticated the dog and cat. Dogs were bred from the fiercest canines—the wolves—and became the hunting partner of humans. Cats, bred from equally fierce small felines, were useful in eliminating vermin inhabiting the grain depositories of early farmers. Grain itself—the very staff of life—has been altered from its natural state by so many generations of controlled genetic manipulation that it almost certainly would not survive well in the wild.

Many organisms in nature have co-evolved with other creatures. For instance, many species of hummingbirds have developed beaks specially equipped to pollinate certain selected species of orchards. And the orchards have expended a great deal of genetic energy making themselves more attractive to their pollinators. So if human activities eliminate a flower, certain bird species may vanish as well.

It should not be believed that all human-caused biodiversity degradation has been due to the actions of evil, selfish, or uncaring people. A laudable goal of human farmers is to grow more crops to feed a growing human population. So farmers have elected over the years to enlist the services of biochemists to develop insecticides such as DDT to protect crops from insect vermin and to control the mosquito population, carriers of the deadly malaria virus. But one reason that DDT was banned was the discovery that, as a side effect, it also weakened the eggshells of certain nautical birds.

Because we have considered human effects upon other species for only a short time, from an evolutionary viewpoint, human carelessness also has an impact. Large cities concentrate certain organisms—such as the rat, roach, and pigeon—at the expense of others. Border fences can disrupt the migratory patterns of many creatures.

The monarch butterfly conducts epic annual migrations from the northeastern United States to its winter home in Mexico. Eco-tourism in the Mexican forests must be rigorously enforced, to make certain that hordes of well-meaning butterfly lovers do not put this species at risk.

Human energy-production systems, even of the most benign forms, can have serious effects upon local ecosystems, if they are designed without adequate foresight. Wind energy is thought of as one of the greenest electric-power production possibilities. But small mammals preyed upon by eagles have learned to hide beneath the spinning blades of California wind turbines, causing mass fatalities among the American national bird's population. Large wind-power farms may also change down-range wind patterns, which can affect the migration patterns of insects and other small organisms.

As developed and developing nations turn toward biofuels such as ethanol to supply transportation fuel and reduce greenhouse-gas emissions, unforeseen effects on the ecosystem can occur. Most biodiversity on planet Earth is found in the rain forest, which also serves as an absorber of carbon dioxide and producer of oxygen. If developing equatorial countries destroy their rain forests to produce biofuel exports, both biodiversity and global climate will suffer.

Even the most noxious, destructive organisms should not be eliminated from the global ecosystem, no matter what our opinions of them might be.* Consider, for example, the termite. Among homeowners reading this book, this wood-munching insect will find few admirers. But in nature, this creature eats and reduces the cellulose in dead wood. If we somehow managed to rid the planet of the termite, forest ecosystems would suffer.

The rain forests are the most vulnerable of the major terrestrial ecosystems, largely because the depth of fertile soil is not large. But these regions are also are home to the most diverse population of species. Although our catalog of rain-forest species is incomplete, many rain forest organisms are known to have significant medicinal properties. It is alarming that unconstrained commerce is destroying rain-forest environments and species before we even understand the resources of these regions.

What Can We Do?

The point of this book is not to predict doomsday, but to help prevent it. If we approach these problems in an enlightened manner, they can be solved.

---

* The authors are willing, however, to cede that some species are best eliminated. We cannot think of a single positive environmental benefit of the smallpox virus.

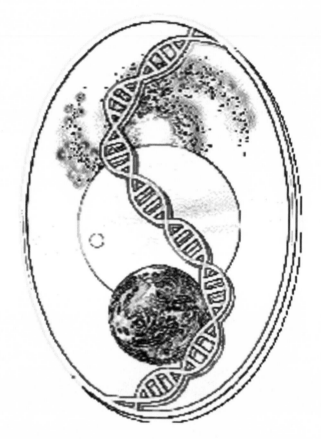

Figure 7.1. An artist's concept of the Earth as an integrated organism, connected by DNA. (Courtesy of NASA.)

First, we might consider the symptoms. Nations might make better use of Earth-resource monitoring technology. Since the problem of ecosphere degradation is global, international data sharing and cooperation are essential. The same space-based technology used to monitor terrorist threats can be applied to protect endangered terrestrial species.

Responsible eco-tourism has already emerged as a positive influence on biosphere protection. Let us hope that it will be expanded during the critical decades ahead.

As we replace and supplement fossil fuels, planning on an international level is required. Biofuel use, wind turbines, and other "green" technologies should be implemented with as much environmental planning as possible. Space research on closed ecological systems may lead to biofuel production modes that have only transient effects upon the rain forest and other threatened terrestrial ecosystems.

But all these approaches treat symptoms. To attack the root causes of the human-caused mass extinction currently under way, human attitudes toward our species and our planet must change. Many consider the best hope to be a widespread adoption of the Gaia hypothesis pioneered by James Lovelock and Lynn Margolis—or at least some aspects of it. Instead of viewing ourselves as separate, disconnected organisms, we might profitably consider each individual life form to be analogous to a living cell and each species to play the role of an organ in an integrated, living planet. Earth might be considered as a living entity, with all of her components linked by the shared DNA molecule (Figure 7.1).

Humans collectively might be considered as the nervous system of this planetary organism. Let us hope that we wisely use our intelligence to nurture and preserve life on our small planet.

Further Reading

An incomparable work on terrestrial biodiversity is David Attenborough, *The Living Planet* (Boston: Little, Brown, 1984). A good reference on the Gaia hypothesis is James E. Lovelock, *The Ages of Gaia* (New York: Oxford University Press, 1988).

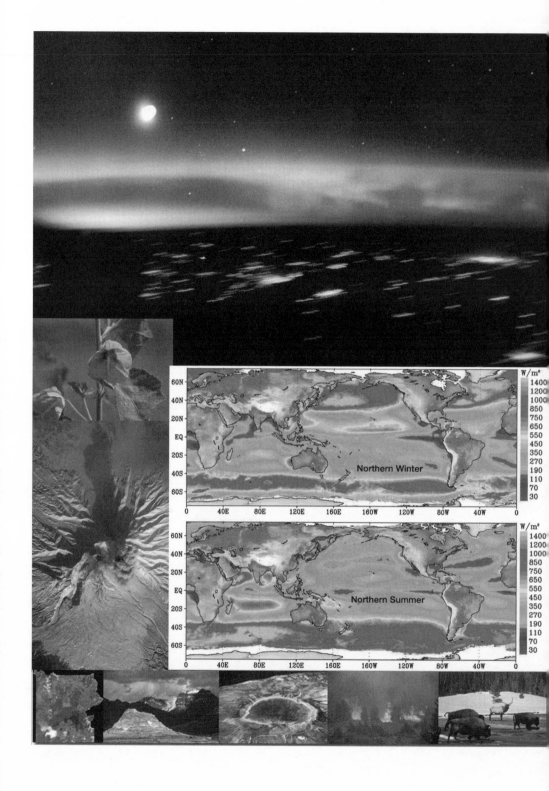

Northern Winter

Northern Summer

W/m²
1400
1200
1000
850
750
650
550
450
350
270
190
110
70
30

Diminishing Energy

As mentioned in Chapter 3, the immutable laws of thermodynamics appear to place a limit on how well we can be stewards of our planet. To recap, the first law states that all energy is conserved. This means that no matter what you do, you cannot get more energy out of a system than you put into it. And some of that energy is inevitably wasted, which is the second law of thermodynamics. Translated into a discussion of global energy supplies, at least when we limit ourselves to discussing the energy problem and planet Earth, these two laws tell us that no matter how clever we may be, no matter how resourceful we become, no matter how hard we try, we will always waste some of the energy we consume.

Where is energy consumed? In 2004, the United States consumed approximately 25 percent of the world's primary energy. About a quarter of that energy was used for transportation, a third was for industrial use, and the rest was fairly evenly split between residential and commercial users.

Other countries use energy too, and as the rapid industrialization of formerly low-energy-consuming countries continues, their need for energy is rising. Almost all of the projected growth is in oil, coal, and natural gas (Fig. 8.1).

Fortunately for our civilization, nature took millions of years to convert the energy received from the sun into plentiful and fairly compact energy storage systems. Oil, coal, and natural gas, the so-called fossil fuels, are most simply understood as batteries, storing energy received by plants and animals from the sun long ago into a compact form that we have learned to use today. Over millions of years, fossil fuels are thought to have formed by

L. Johnson et al., *Paradise Regained*, DOI 10.1007/978-0-387-79986-5_8,
© Praxis Publishing Ltd, 2010

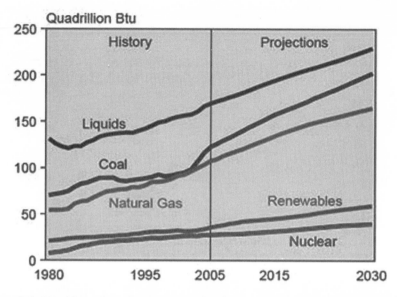

Figure 8.1. World marketed energy use by fuel type as projected by the United States Energy Information Administration. (From U.S. Energy Information Administration [EIA], International Energy Annual 2005. Washington, DC: EIA, June–October 2007.)

the action of heat from Earth's core and pressure from rock and soil on the remains, or fossils, of dead plants and animals. When you drive your car or turn on your cook stove, you are essentially using energy from the sun that fell upon Earth in the distant past and was stored until it was processed and piped to your car or home. Unfortunately, the world's supply of these resources is finite and we will eventually run out.

There is some debate as to when production of oil, coal, and natural gas will peak, and when our reserves will be expended. The date when the well will run dry for each of these depends on several factors: the rate at which they are used, the amount that remains to be used in known or yet-to-be discovered locations, and how serious we become about conserving the fuel for later generations. But one thing is certain: these fuels will become increasingly scarce and expensive. The only open question is when. Whether it will be in 50 years, 100 years, or 300 years is a matter of mostly academic debate. But that date will eventually arrive and we had better have a backup plan. What are the options?

Figure 8.2. The Brown's Ferry nuclear power plant near Athens, Alabama. (Courtesy of the United State Nuclear Regulatory Commission.)

Nuclear Power (Fission) (Fig. 8.2)

The United States, France, and Japan together produce approximately 57 percent of the world's nuclear generated electricity. In the United States, 19 percent of the electricity consumed comes from nuclear power. In France, 78 percent of electrical power is nuclear. Nuclear energy is not somehow directly extracted from an atom and turned into electricity. Rather, the process by which power is generated from a nuclear source resembles how power is produced in a conventional coal, oil, or natural gas fired plant—through the generation of heat. In this case, splitting uranium atoms in a process called nuclear fission produces the heat. When a specific target atom is struck by a neutron, it forms two or more smaller atoms releasing energy and, when properly configured, more neutrons. These neutrons strike other atoms, releasing yet more neutrons and more heat. The heat is used to boil water and produce steam, which then drives a turbine to make electricity.

The fuel for the nuclear reactor is usually the element uranium. Uranium is fairly common, traces of which can be found in most rocks, dirt, and in the oceans. This uranium can be processed for use in nuclear power plants directly, or it can be used in another type of reactor that makes its own fuel. The latter is called a breeder reactor because during the nuclear chain reaction process it "breeds" fuel for future use. With known uranium reserves and breeder reactor technology, nuclear power could provide humanity with abundant electricity for millennia. It is also important to note that generating electricity in a nuclear power plant does not produce any greenhouse gases.

Nuclear power is not without its critics, and some of the concerns raised are serious. The safe operation and eventual disposal of nuclear power stations is of paramount importance. Even a small radiation leak is environmentally unacceptable and a large leak could prove catastrophic. Fortunately, the accident at Three Mile Island in the United States killed no one and produced no environmental damage. The same cannot be said about Russia's Chernobyl accident. Chernobyl was by far the worst reactor accident in history, producing both loss of life and localized environmental damage. The accident resulted from the reactor operators ignoring their own rules, disabling key safety features that would have otherwise prevented the accident. Above all, the reactor had no containment vessel to trap the radioactive gases that were released—a required feature of virtually all commercial reactors used to generate electricity in the rest of the world.

The biggest drawback to nuclear power, in our opinion, is the safe storage of the radioactive waste generated as the by-product of a reactor's operation. The United States has no active central repository for this deadly waste, which can remain toxic for many thousands of years. Much of the generated waste is in temporary storage at the plants themselves awaiting shipment to a more permanent storage facility elsewhere. Unfortunately, this temporary storage is looking more and more like a permanent situation and is certainly not a viable permanent solution. Susceptible to accident or deliberate tampering, this on-site stored waste is a relatively near-term problem that must be fixed.

Nuclear power plants can, in the wrong hands, be used to produce fuel for nuclear weapons. A serious risk of the technology is that its stated peaceful purpose can fairly easily be diverted to making bombs and increasing the risk of nuclear war occurring somewhere in the world. Having this technology in use by unstable or rogue states could prove to be a gross mistake.

Another drawback is the sheer scale at which nuclear reactors would have to be built in order to replace power plants currently using oil, coal, and natural gas. Without increased efficiency and advanced grid management methods, thousands of new nuclear power plants will be required just to meet existing demand. The risk from diversion of nuclear fuel to terrorists, from an accident, or from misuse of reactor technology by countries seeking to develop nuclear weapons makes nuclear power a potential global energy solution that should be undertaken only when most, if not all, other options have been ruled out. The key word in this assertion is the word *global*. Unquestionably, the United States, Europe, Japan, as well as other stable and peaceful governments should be building nuclear power plants now to meet our near-term energy needs (over the next 50 years), but they should only be considered an interim solution.

Nuclear Power (Fusion)

There is a joke among physicists: "Fusion is the power source of the future—and it always will be." (There is an alternative version: "Fusion power is 30 years away—and always will be.") This appears to be the unfortunate reality of fusion energy research for the last 50 years, and, barring a breakthrough, the trend seems likely to continue. What is nuclear fusion and why does it appear to be so hard to use it for power production?

The energy produced in the sun comes from nuclear fusion. In the sun, hydrogen atoms are being tightly compressed and fused together to form helium. This process has been ongoing for 4.5 billion years and will continue for at least a few billion more. Creating a fusion reaction in the laboratory is not the problem; that is something scientists began doing in the middle of the last century. The problems are with control and something called "breakeven."

An uncontrolled nuclear fusion reaction is a bomb. The hydrogen bomb uses nuclear fusion to produce tremendous devastation and should not be confused with the fission bombs dropped on Japan during World War II. A fusion bomb uses the energy produced in a fission bomb to get the reaction started, whereupon an uncontrolled fusion reaction ensues, releasing significantly more energy than is possible with a fission bomb alone. Even the most primitive fusion bombs produce about five hundred times as much energy as a fission bomb.

The fusion reaction can be controlled, and this too has been routinely achieved in the laboratory. The problem is that, for now, it takes as much or more energy to create and sustain a fusion reaction than it produces. As long as you have to put more energy into fusion than you get out of it, fusion will not be a viable (net) energy producing process. The point at which you are able to extract as much energy as you put into a fusion reaction is called "breakeven."

Finding a consistent, economically viable way to reproduce the conditions in the sun to produce fusion has been the challenge; the solution, at least at the current pace of research, may be just "30 years away." However, if the resources of the solar system are brought to bear, specifically the vast stores of helium-3 embedded in the lunar regolith, then nuclear fusion may not be so far away after all. (See Chapter 10 for more information about helium-3.)

Hydropower (Fig. 8.3)

Another source of electricity that produces no direct greenhouse gases is hydroelectric power. This power comes from the use of flowing water, typically produced when a dam is constructed across a river. Instead of steam, as is used in a fossil fuel or nuclear power plant, hydropower uses falling water to turn a turbine and produce electrical power. The process is relatively clean and efficient. However, it is not without environmental consequences. When a dam is constructed, a river is usually turned into a lake. The lake encompasses land that used to be available for other purposes: homes, farmland, industry, or, at the very least, as a habitat for various land

Figure 8.3. The Lake Guntersville Dam and hydroelectric power plant near Guntersville, Alabama. (Courtesy of the Marshall County Alabama Convention and Visitors Bureau.)

animals. All of these uses are precluded when the area is flooded to create the lake that then provides the water for generating hydropower.

Many of the power-producing dams in the United States were built in the 1930's and 1940's, long before our current age of environmental awareness. The environmental impact of building a single dam, let alone the many thousands that would be required to make a dent in our dependence on fossil fuels, would be unacceptable. Also, there may not be enough suitable locations for building these dams, even if the local environmental consequences were considered not to play a major role. All the easy sites are taken.

Another form of hydropower is tidal power generation. Harnessing the tides to turn turbines and generate electricity is technologically feasible and has been demonstrated. Unfortunately, tidal power stations can produce power only along the cycle of the tides (peaking every 6 hours), which is not likely to track the human power demand cycle. There is also the open question of what these large underwater turbines will do to the coastal environment and its aquatic residents. Hydropower may provide part of a global solution but, at best, only a small part.

Wind Power (Fig. 8.4)

In a fossil fuel or nuclear plant, steam is used to turn a turbine and generate electricity, so why not use wind power? The idea is not new, and several wind farms are now populated with multiple windmills and regularly generating electricity. Currently, about 1 percent of the world's electrical power is produced using wind. Wind energy is plentiful, renewable, clean, and free of greenhouse gas emissions. There is no technical reason why many thousands of windmills cannot be built and added to the power generation infrastructure today. But there are technical reasons why they cannot be the solution to the world's energy needs.

The first problem is one of wind availability. The wind is not constant and it does not increase when the power demand is high. Is it practical to have a power grid that has to shut down, or perhaps just "brown out," when the winds shift? Also, the amount of power generated is highly dependent on the speed of the wind. In studying one windmill farm, engineers found that about half the total energy available was produced in about 15 percent of the operating time. The rest of the time the power output was low, reducing the amount available to consumers.

The other major problems facing wind power are environmental. Some of the best locations in the world for wind power generation are right in the

Figure 8.4. The Buffalo Mountain, Tennessee, Wind Park is the largest in the southeastern United States. (Courtesy of the Tennessee Valley Authority.)

middle of protected lands or on the horizon of majestic landscapes where people are loath to allow large man-made structures to mar the view. Do we want to cover our public lands with huge towers and power lines? Or the oceans just off the coast of the northeastern United States?

Another environmental problem is the scale of the land use required to generate large quantities of power. For example, a 200-megawatt wind farm requires approximately 20 square kilometers of land. This is mainly due to the spacing required between each windmill so as to reduce the power generation losses that result when they are too close together. Fossil fuel plants with much higher power output require significantly less land.

Wind power is another good idea for supplementing the energy supply, but it is not a viable long-term solution.

Figure 8.5. Geothermal energy sometimes may be readily accessible, as seen in this photograph of the Old Faithful geyser in Yellowstone National Park. But it is the very fact that geothermal energy is available in only a few places that make it a popular tourist attraction to begin with. (Courtesy of the United States Geological Survey.)

Geothermal Energy

Another niche power production system uses Earth's natural geothermal energy. In various parts of the world, steam generated deep within Earth escapes to the surface. The resulting hot springs have been used for centuries for bathing, heating, and as cures for various maladies. They are also popular tourist spots. The Old Faithful geyser in the Yellowstone National Park is perhaps the most well known (Fig. 8.5). Early in the 20th century, the first experiments were conducted in using this natural form of steam to generate electricity—and it works. Unlike most other forms of renewable energy such as wind and solar, which are fundamentally intermittent, geothermal can generate base load power.

Across the globe entrepreneurs are tapping into geothermal energy to build power plants and generate electricity in a manner that does not pollute the atmosphere. Unfortunately, the very geologic formations that make this power source possible also may place limitations on its overall viability. In many cases, outside water must be pumped into Earth's "hot spots" to make steam and produce power. The water is then pumped elsewhere to cool. This process sometimes makes the nearby soil unstable, producing cave-ins and landslides, and sometimes small earthquakes.

The aforementioned concerns will limit how widespread the technology may be used. Conventional geothermal energy is available in only a few locations (where Earth has provided access to its deeply produced, innermost heat) and only some of those are sufficiently stable so as to allow building large-scale power plants. Elsewhere, realizing geothermal potential requires costly engineering intervention.

Like wind energy, geothermal energy is an excellent supplemental energy source, but it is not scalable to meet the needs of a global, high-technology civilization.

Biofuel

A lot of attention has been paid to biofuels in the last few years. Basically, a biofuel is one that is derived from recently dead plants (as opposed to fossil fuels, whose energy is derived from long-ago deceased plants). The most common biofuel is alcohol generated from corn or sugar cane. With the increasing price of oil, making gasoline with 10 to 20 percent biofuel alcohol has become cost competitive.

There are serious questions about the overall energy efficiency of the process, given that the plants must be planted and harvested using machines, processed by machines, and then distributed (again, by machines) to where the users may be located. Taking into account the energy required in each of these steps prior to consumers filling up their tank, some scientists doubt that the energy output exceeds the energy input. In fact, it is possible that more energy will be consumed in making a liter of biofuel than will be extracted from it by our cars and trucks.

There is also the problem that raising the huge amount of plants that will be required to sustain a viable biofuel infrastructure may pit the use of arable land for use in cultivating biofuel against that same land being used to grow food. Do we really want to have food and fuel competing against one another? We think not, especially in a world that still sees a large fraction of its population going to bed hungry each night.

Conclusion

None of the alternative energy sources described above is a bad idea. None should be dismissed as part of a global renewable and green energy strategy. And none of them individually, nor taken together, provides a viable long-term solution to the world's increasing energy needs. Something new and different must be tried, such as getting power from space, which we discuss in Chapter 11.

Humans Before the Industrial Age: A Desirable Ecological Goal?

We are the music-makers,
And we are the dreamers of dreams,
Wandering by lone sea-breakers,
And sitting by desolate streams;
World-losers and world-forsakers,
On whom the pale moon gleams:
Yet we are the movers and shakers
Of the world, forever, it seems.

Arthur O'Shaughnessy, from "Ode"

All of us have experienced the darkness and felt despair. Our civilization's problems seem beyond measure—an exploding population, pollution, energy shortages, climate change, terrorism, and nuclear proliferation are among those threats. There is nowhere in space to flee to; Earth is filled up, the moon may be waterless, Mars is a lifeless desert, and the stars are just too distant.

But if we cannot flee to greener pastures, perhaps we can maximize our chances of surviving a future global catastrophe. Then we can emerge from our holes and help direct the surviving human remnant on a different, simpler path. Some believe that if we forsake the world of the cities, cell phones, cars, and computers, then we can return to a simpler, sustainable preindustrial golden age. Has such an age ever existed? What would be the consequences if we retreated from our technologies?

The Paleolithic Era: Life in the Old Stone Age

Let's first imagine that we chuck it all. We reset the human clock as far back as possible and return to the good old days of the Paleolithic Era. Population was much smaller and human-caused pollution nonexistent in this, the

L. Johnson et al., *Paradise Regained*, DOI 10.1007/978-0-387-79986-5_9,
© Praxis Publishing Ltd, 2010

longest period of human prehistory, which extends from about 2,000,000 BC to 10,000 BC. We could all be movers and shakers then, since governments and bureaucratic structures did not yet exist. But our lives would not be very long or very interesting.

Most human activity in this period centered on the quest for food. Since we did not farm or know how to maintain animal herds, it was necessary to gather nutritious herbs, follow game animals, and opportunistically steal the kills of larger predators.

This must have been a lot harder early in the Paleolithic Era since fire had not yet been tamed and our best cutting tools were made of hand-carved flint and rock. Until fire was mastered about one million years ago, our ancestors must have consumed their food raw. As well as lacking in taste, raw meat often contains various parasites. This fact contributed to the short life expectancy of our early ancestors. We would have started having babies in our early teens. By the time those fortunate survivors reached the age of 30, they would have been considered the elders of the tribe.

Medicine did not exist, nor did dentistry. With no knowledge of sanitation, many women and infants must have died in childbirth. Our Paleolithic forebears may have compensated for this by developing and orally passing on knowledge regarding the effects of various beneficial herbs.

Even when Paleolithic hominids succeeded in finding ample food sources, they were faced with a formidable problem: locating shelter. Without sophisticated metal tools, building anything like a modern house would be out of the question. The most sophisticated human dwellings from the late Paleolithic Era were houses made from animal bones and antlers. Perhaps animal hides, which have not survived in the fossil record, were used to insulate these shelters as they were used to clothe our ancestors.

Earlier, human bands must have competed for the most suitable caves in mountainous regions. Not only would our ancestors contend with other human bands and our Neanderthal cousins for such choice habitats, they also competed with such formidable cave-loving creatures as bears and huge felines. It was inevitable that humans were not always victorious in these contests.

Then there was the small matter of mobility and tribe size. Draft animals were far in the future; all land travel was by foot. If you could not get along with the few dozen people in your band, tough luck! The nearest neighboring tribe might be far beyond your walking range.

Let's Try the New Stone Age: The Neolithic Era

It is unlikely that we would find Paleolithic life enticing. So let's skip forward a few millennia to the New Stone Age. This Neolithic Era extended from about 10,000 BC to 4,000 BC. During this era, at least some people lived in mud-brick houses. Although these were not equipped with indoor plumbing and lighting, they at least kept the elements out. Neolithic food was a bit better and more dependable than Paleolithic food. Agriculture was being developed and some crops were farmed. Instead of chasing fleet game animals with spears and bows, our Neolithic ancestors could choose a tender morsel from the domesticated beasts and fowl.

But sanitation was primitive. The odors from dumped slops in Neolithic towns such as Catal Huyuk and Jericho must have been intense as local population levels exceeded a thousand.

Medicine and dentistry had not advanced. The human life span had not substantially improved from Paleolithic levels. And there was a new wrinkle to add even more difficulty for the human participants: organized warfare had been initiated. Probably, the first wars were not between rival agricultural settlements. Instead, the comparatively rich lifestyles of the agriculturists and animal herdsmen must have attracted the unwanted attention of nomadic Paleolithic tribes. Early on, Neolithic towns constructed defensive walls to keep the nomads at bay.

But human land mobility had increased since some domesticated creatures—notably the ox—could be used as draft animals. During the late Neolithic Era, humans may have used the recently developed sail to venture out on the open seas. In the eastern Mediterranean, village culture began to spread from the mainland to offshore islands.

Although Neolithic humans ate better and lived more securely than their Paleolithic cousins, life was still harsh and short, and farming and building were done with only primitive stone tools. So let's jump forward a bit further in time from the Neolithic dawn.

The Bronze Age: Civilization is Born!

About 6000 years ago, some unknown genius discovered that certain metals could be processed and altered with the aid of fire. With this bit of mastery over the physical world, numerous possibilities arose for our ancestors. Villages grew into cities. With the use of metal tools, hydraulic systems were developed to supply and store fresh water and to remove sewage. In some of the new cities, these advances led to populations of ten thousand people or more.

Humans ate better as metal tools allowed farmers to till the land and harvest crops with greater efficiency. The new technologies allowed for the construction of impressive, multistoried buildings.

Writing was developed, which led to legal codes and centralized governments. In the new cities, various social classes arose. You would do well if you were born into the ruling caste, but not so well if you started your life as a slave.

With their newfound powers, responsible rulers strove to distribute agricultural produce fairly among the populace and to store food against the famines and droughts that would surely follow the abundant "seven good years." For the first time, some people had the leisure to think about the meaning of existence, the afterlife, and other related topics. Religious cults of Earth goddesses and sky gods gained followers in many parts of the world. Human sacrifice and ritual self-mutilation were not unknown.

In certain regions, notably the Aegean, a golden age of relative peace and prosperity occurred. But it should not be supposed that such an occurrence was due to improvements in human nature. In all likelihood, the naval technologies of the Cycladic Islands temporarily protected local populations from the depredations of marauding armies equipped with chariots drawn by the newly domesticated horse.

As population increased, civilization began to spread from its original birthplaces. In some regions, city-states joined to form the world's first empires. To provide hot water to increasing populations, many Bronze Age towns were constructed in tectonically active regions—not a very good idea from our perspective!

For the rich, at least, medical care began to improve, as written compilations of medical practice allowed for the training of new generations of practitioners. But the average life expectancy was still far below what we enjoy today.

The Bronze Age was a time of adventure and expansion. But for most people it was a difficult life.

The Age of Iron

About 3000 years ago, people mastered the use of iron. Applied as weapons, this strong metal led to the growth of organized armies. Empires spread across the globe and competed with each other. In the west, Persia was supplanted by Athens, which itself was ultimately absorbed into Rome. Across the tortuous Silk Road, trade flourished between the Romans and the growing eastern empire of China.

A network of roads was constructed, and cities that could house 100,000 people or more were built. It seems unlikely that the innovations of the earlier hydraulic engineers of the Bronze-Age Aegean could keep pace with the population growth. Perhaps because of both population growth and poor sanitation, plagues were not uncommon. It was much more likely that a random person would die horribly from a plague than as a result of the incessant warfare.

The old gods and goddesses had failed. They could prevent plagues and wars no better than they could forestall earthquakes and volcanic eruptions. Faced with the choices of mindlessly worshiping these ancient deities or participating in the mind-drugging, bloody circuses supplied by the ruling elites, it is not surprising that many elected to become monotheists. The spread of religion and ethical philosophy in this period was greatly influenced by the adoption after 1000 BC of alphabetic scripts, which greatly enhanced literacy. Art and literature flourished as well—at least for the affluent. It is difficult to imagine that the average person would have had time for such luxuries.

Ruins dated to the Iron Age abound in Western Europe. In some places, buildings, highways, and aqueducts constructed in that period are still in use.

The Middle Ages: A Time of Turmoil

The cause might have been the failure of the ancient gods or the successful incursions of the barbarian hordes. Perhaps climate change played a role as crop failures made organized military adventures more difficult. Whatever the ultimate reasons, the organized bureaucratic structures of the unified Roman world began to decay and disappear before 500 AD.

Population plummeted and literacy declined in Western Europe. Where peaceful farmers had gathered the fruits of the field, and where philosophers and scholars had debated ethics and physics, scattered bands of survivors struggled against each other and against the elements in a world without large-scale organization.

One of the surviving fruits of the late classical world was monotheism. If western civilization had not collapsed when it did, monotheism might have become a unifying force. Sadly, this was not to be the case. A widespread belief in one creator soon degenerated into a series of competing branches of monotheism, each maintaining the same sacred scripture, but each with its separate interpretation. Not only did Christians, Jews, and Moslems fight with each other; vast quantities of blood were also spilled in strife among rival sects of Christianity and Islam.

One thing that united the medieval monotheists, however, was a hatred of the polytheistic religions that had preceded them. Woe to the independent woman who helped preserve some ancient knowledge of herbal medicine; she was likely to be declared a witch and burned at the stake!

With the primitive sanitation and medical practice, plagues caused by human stupidity and superstition ravaged the population. But to be fair, there were some bright spots in the midst of this medieval murder and mayhem. Visual art had originated during the Paleolithic Era as a form of sympathetic magic. Cave paintings of mammoths and other large herbivores were then thought to increase the herds and the hunters' yield. Painting and sculpture developed further during the Bronze and Iron Ages, resulting in the masterpieces found in the collections of the world's great museums. But in the Middle Ages, when illiteracy was widespread and the new religions required visual methods to teach scripture to the masses, painting and related arts had a significant flowering. Many of these works have survived and can be enjoyed on the walls of cathedrals, mosques, and synagogues constructed during that period.

The visual arts were also used to illuminate hand-copied classical manuscripts that had survived the fall of Rome. We owe a great deal to the scribes and manuscript illuminators of that era.

Music also advanced during the Middle Ages, perhaps for the same reasons as the visual arts. Choirs perform hymns composed by such notables as Hildegard of Bingen to this day.

If you were very lucky during the Medieval Era, you might have worked as an architect, artist, or composer. Or you might have led a sheltered life as a scribe in a protected religious enclosure. But for most people in that period, life was squalid, uncertain, and short.

The Renaissance: Civilization Awakens

Around 1400 AD, the European Renaissance began, and it is widely viewed as a reawakening and revival. Perhaps because of the recently developed printing press, more books were published. Literacy and learning increased after languishing for a millennium. The practice of modern science began. Humans learned that their world circled the sun and was not the center of the universe. Scientific principles—anatomy, the camera obscura, and perspective—were applied by prolific painters and sculptors to their rapidly evolving disciplines.

In both the West and Far East, the technology of the sailing ship reached new levels of perfection. Trade, exploration, and colonization became global enterprises.

But much of this advance came with an increase of international turmoil. If the Ottoman Turkish conquest of the remains of the ancient Eastern Roman Empire had not precipitated a large-scale migration of eastern scholars to the Italian city-states, the renaissance may not have happened. The schism of western Christianity into Protestantism and Roman Catholicism resulted in some of the bloodiest wars that humanity had known.

Was There Ever a Golden Age?

Our search through history for a golden age comes up empty. In every previous era, things might have been exciting and glorious for the lucky few. But for the majority in most past eras, life was more uncertain than it is today.

We should learn from the past. We should apply the wisdom of our ancestors and avoid their follies. But we should never forget that humans are the "movers and shakers" of Earth. It is our obligation to look toward the future and apply our arts and sciences to ensure a good life for all the peoples of our emerging global civilization.

Any attempt to restore our planet's ecology that ignores the human element will fail. If the majority of people are poverty stricken or victimized, ecological awareness will fail. Since ours is the first era in which a substantial fraction of the human population lives well, ours is the first with a chance to restore the planet. Let's rise to the challenge!

Further Reading

Many authors have considered the laborious development of human culture and civilization, the challenges and capabilities of our ancestors. For a survey of prehistory and history through the Iron Age, see Jacquetta Hawkes, *The Atlas of Early Man* (New York: St. Martin's Press, 1976). Another wonderful source is Jacob Brownowski, *The Ascent of Man* (Boston: Little, Brown, 1973). Finally, no student of civilization's development can ignore Kenneth Clark, *Civilization* (New York: Harper & Row, 1969).

Paradise Regained

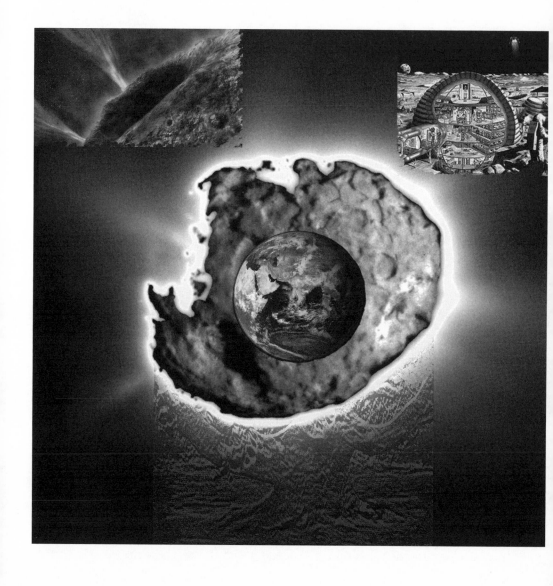

Raw Materials from Space

10

Allons! We must not stop here,
However sweet these laid-up stores, however convenient this
  dwelling we cannot remain here,
However sheltered this port and however calm these waters
  we must not anchor here,
However welcome the hospitality that surrounds us we are
  permitted to receive it but a little while.

Walt Whitman, from "Song of the Open Road"

Walt Whitman was right: Earth is not a closed system. The road to the riches of the solar system lies open. Nature has provided humanity with a virtually unlimited supply of raw materials, and that supply is right above our heads. The asteroids, comets, planets, and moons of the solar system contain enough of the soon-to-be-scarce raw materials required to feed our technological civilization for thousands of years.

As the nations of Earth increase their consumption of raw materials to supply our expanding global civilization, prices will inevitably rise due to the economic laws governing supply and demand. Earth has a finite supply of fossil fuels, copper, zinc, titanium, and other elements as well as the minerals composed of them. In a peaceful world, competition for these resources will lead to higher prices, lower consumption, and recycling of existing supplies. If these approaches fail, as they often do, armed conflict may result. Wars have been fought over resources in the past, as they will be in the future. In an age where many peoples have weapons of mass destruction, the consequences are unthinkable.

What are the resources of the solar system? Where do we find them? How do we retrieve them? What do we wish to do with them? These major issues are addressed in this chapter.

L. Johnson et al., *Paradise Regained*, DOI 10.1007/978-0-387-79986-5_10,
© Praxis Publishing Ltd, 2010

An Inventory of Solar Riches

In considering the riches of the solar system, it is convenient to divide it into a number of discrete regions, not too dissimilar to the divisions of the living Earth into various discrete biospheres.

Near the Sun

First, we discuss the sun, which may be considered as a "zone of fire." Spewing from the sun into the interplanetary medium is the highly variable solar wind. Moving at 200 to 800 kilometers per second, this stream of ionized hydrogen and helium nuclei and electrons has a typical (but highly variable) density of ten particles per cubic centimeter, at Earth's solar orbit.

Space miners might consider using hydrogen from the solar wind to produce water (in combination with oxygen) for use by lunar colonists. A rare form of helium—helium-3—is also present in the solar wind and could theoretically be retrieved for use in terrestrial fusion reactors.

But there are problems. To mine the solar wind on a practical scale, one requires an electromagnetic scoop about 1000 km in diameter. Although the superconducting materials required to maintain scoop function work well far from the sun, today's superconductors will certainly fail in the inner solar system, where temperatures and solar-wind densities are high. So barring breakthroughs in material science, direct mining of the solar wind remains an option for the far future.

Moving out from the central solar fire, we next come to the realm of the inner planets Mercury and Venus. At one-third Earth's solar distance, Mercury is a challenging destination for contemporary spacecraft. Multiple trajectory-altering passes of Earth, the moon, and Venus is one way to project a spacecraft on a Mercury-bound path. Once we have arrived, are there any useful Mercurian resources? Surprisingly, the answer to this question seems to be yes. Billions of years ago, water-bearing comets repeatedly crossed the inner solar system, sometimes impacting the inner planets. Even though conditions on small, airless, and hot Mercury do not seem ideal for water retention, Earth-based radar studies reveal water-ice deposits near the planet's poles. These are probably comet-impact remnants in the interior of craters protected from sunlight. Such water may be of interest to an eventual solar-system civilization, but seems of little importance to water-rich Earth.

We will skip torrid, greenhouse Venus in our resource inventory. Resources may certainly be present there, but retrieving them from this inhospitable world is problematical.

Earth's Moon

Humans have visited in person only one solar-system world beyond Earth: our planet's large moon. At present, many space-faring nations or transnational entities—the United States, Russia, Europe, Japan, China, and India—are enhancing their capabilities so that astronauts can again visit the moon in the decade between the years 2020 and 2030. Certainly national pride is a factor here, and we can expect flags of many nations to be raised above the silent lunar surface. But it is possible that the new moon efforts will also contribute something to terrestrial civilization.

Katharina Lodders and Bruce Fegley, Jr. tabulated estimates of lunar composition, largely based on analysis of lunar samples returned by Apollo spacecraft. The moon's mass is about 44 percent oxygen, 19 percent magnesium, 6 percent aluminum, 20 percent silicon, 7 percent calcium, and 3 percent iron, with trace amounts of other elements.

Since the moon and Mercury are similar in many respects, it is hoped by many that deposits of water ice will be confirmed in lunar craters near the poles, protected from direct sunlight. At present, the evidence for such a resource is ambiguous at best.

As tabulated above, the moon is approximately 20 percent silicon by mass. By happy coincidence, silicon is the major element in certain varieties of photovoltaic cells. So it is not impossible that future astronauts could mine silicon from the lunar crust to construct giant arrays of photovoltaic cells (Fig. 10.1). This energy could be converted to microwaves and beamed back to Earth to supply a substantial fraction of our planet's energy needs.

Sun power from space is discussed in greater detail in Chapter 11. It is enough here to mention that large craters near the lunar limbs could be covered with thin-film silicon photovoltaic cells. If care was taken, the lunar face would not be greatly altered by this process.

Solar energy would first be converted into electricity and then into microwaves. These microwaves could either be transmitted directly to Earth receiving stations or to relay satellites near the moon. Conceivably, a lunar base or colony could support itself by selling solar power to energy-hungry terrestrials.

Deep within the sun, nuclear processes combine hydrogen atoms to form helium, releasing enormous energy in the process, providing the power that keeps the sun shining and Earth warm. Some of this helium is in a rare form called helium-3. Helium-3 contains two protons and one neutron rather than the usual two neutrons. While rare on Earth, helium-3 should be rather plentifully embedded within the surface dirt, or regolith, on the moon after its bombardment for millions of years by the solar wind.

Why is this important? Helium-3 might be the nuclear fuel we need to

Figure 10.1. Artist's conception of a solar-powered lunar mining facility. (Courtesy of NASA.)

make mass-scale nuclear fusion reactors possible, providing relatively clean and safe nuclear power to future generations on Earth. As stated in Chapter 8, electrical power generated by break-even nuclear fusion is years away from viability, and may always remain so unless something comes along to change the way we are approaching the technical problems. Helium-3–based fusion might be the disruptive technology we need to make fusion power viable.

It is easy to demonstrate that only a small amount of helium-3 would be enough to supply our civilization's entire energy requirement. Consider a scenario in which we require globally 30 trillion (3×10^{13}) watts of electric power. Furthermore, assume that this is all to be supplied by fusion reactors

Frontispiece from
the Introduction

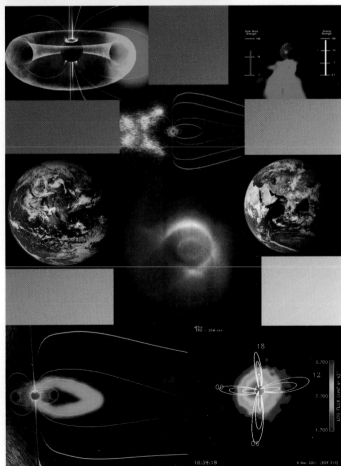

Frontispiece from
Chapter 1

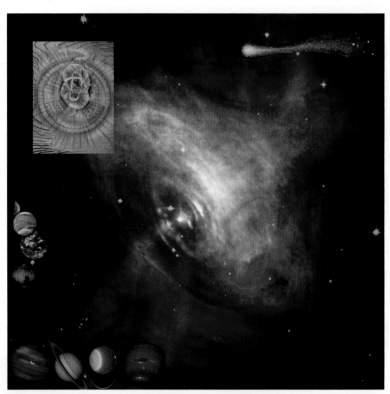

Frontispiece from Chapter 2

Frontispiece from Chapter 4

Frontispiece from Chapter 5

Frontispiece from Chapter 6

Frontispiece from Chapter 7

Frontispiece from Chapter 9

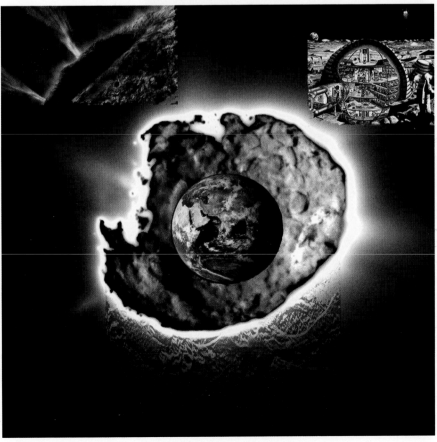

Frontispiece from Chapter 10

Figure 12.2. These pictures, taken in 1973 and 2003, respectively, show how civilization has consumed forestland in Paraguay and Brazil. The Iguazu National Park in Argentina, which is protected from development, remained intact. (Courtesy of the United Nations Environmental Program.)

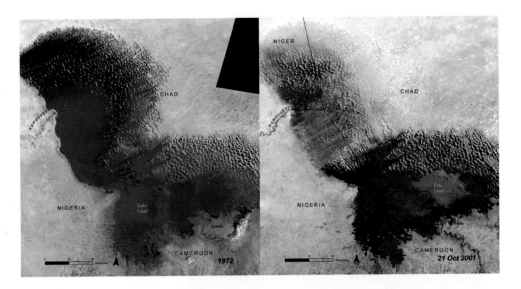

Figure 12.5. Satellite images of Africa's Chad Lake showing a dramatic decrease in the lake's water level (blue) between 1972 and 2001. (Courtesy of the United Nations Environmental Program.)

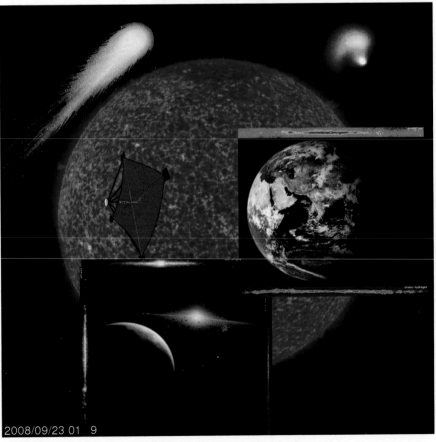

Frontispiece from Chapter 13

Frontispiece from
Chapter 15

Frontispiece from Chapter 16

that burn a mixture of heavy hydrogen (deuterium) and helium-3. Advantages of this thermonuclear fuel mixture include relatively easy ignition and low waste radiation. Every year, about 10^{21} joules of fusion energy (E) are required. If we next apply Einstein's famous mass-energy conversion equation, $E = \varepsilon m c^2$ (where ε is the fraction of reacting mass m converted into energy in our fusion reactor, and c is the speed of light) for the case of thermonuclear fusion, which converts a maximum of 0.4 percent of reactant mass to useful energy, we find that only about 1 million kilograms per year of helium-3 is required by a reactor technology burning 50 percent helium-3 by mass.

Current fusion reactor research is primarily concentrated on the combining of hydrogen atoms to produce helium, releasing energy in the process. The energy would then be converted into electricity. The reactions currently in development release a significant amount of their energy in the form of radioactive neutrons. These neutrons pose a significant health hazard and must be managed accordingly. In contrast, a fusion reaction involving helium-3 and hydrogen would produce very little residual radioactivity. It is possible to envision hundreds of relatively clean fusion reactors providing electrical power for our cities with their fuel being imported from the moon.

Current estimates place the amount of helium-3 on the moon at about a million tons. This should be enough to meet the world's energy demands for thousands of years, even if we do not develop additional alternative clean energy sources. Unfortunately, obtaining the helium-3 from the moon may not be easy. Since the element is embedded in the lunar regolith, the surface dirt on the moon will have to be strip mined and processed to release the helium. To accomplish this, an infrastructure will have to be erected there to allow the mining, transport, heating, and redistribution of tons of surface dirt.

Unlike on Earth where strip mining permanently mars the surface and destroys local ecologies, the impact on the moon will be extremely minor. The moon's surface is already dead, scoured of any potential life by the relentless solar radiation baking it for 14 days at a time. Gathering up this dirt and then redistributing it will damage no ecologies, for there are none there to damage. Using this resource will help preserve ecologies here on Earth.

But there is one catch to lunar mining and settlement plans. (In space, there is almost always a catch or two!). An omnipresent layer of dust bothered Apollo astronauts trudging or riding across the lunar surface. Remote videos of Apollo lunar module liftoff from the lunar surface observed that rocket blasts blew up so much dust that the flags planted by astronauts actually waved in the lunar vacuum. This material coated the astronauts' spacesuits. The interior of the lunar modules became dust repositories. Future lunar colonists and miners will find this dust layer to be a source of inconvenience and frustration.

Compounding the problem is the low lunar surface gravity. Large-scale lunar mining may raise lots of dust, and because of the low gravity, lunar dust will not settle rapidly. Not only would uncontrolled lunar mining raise a dust layer inconvenient to terrestrial lovers of the moon; it might negate plans of astronomers to operate observatories on the lunar far side. Clearly, lunar development must be carefully thought out. But beyond the moon and well within reach of planned interplanetary spacecraft is another source of extraterrestrial resources.

Near-Earth Objects: If We Gotta Move Them, Why Not Use Them?

Between Venus and Mars, there is a class of rocky, stony, and icy objects that orbit the sun. Some are the size of boulders, and others are as large as mountains (Fig. 10.2). These near-Earth objects (NEOs) are either asteroids or extinct comets. As will be discussed in Chapter 13, they occasionally whack the Earth with devastating consequences. To protect our planet and ourselves, we must track and categorize these NEOs. Then we must investigate and develop techniques to divert the trajectories of those that threaten Earth. But if we have to move them, why not use them?

Figure 10.2. Asteroid Eros, a large near-Earth object. (Courtesy of NASA.)

Spacecraft have visited asteroids before, and from each mission we have learned a great deal about their composition. The NASA Galileo spacecraft imaged asteroid 951 Gaspra in 1991 and asteroid 243 Ida 3 years later. In 1997, the NASA Near-Earth Asteroid Rendezvous (NEAR) mission captured images of 253 Mathilde and visited asteroid Eros in 1999, ending its mission by gently crashing into Eros's surface. The Japanese explored the asteroid 25143 Itokawa with their Hayabusa spacecraft in 2005.

The vast majority of asteroids (93 percent) are composed of stone, and the rest are made of iron, nickel, and other metals, or a mixture of these metals with stone. Billions of tons of raw materials are hanging directly above our heads.

According to the Roskill Information Services, world production of iron ore exceeded 1.4 billion tons in 2006. The rapidly expanding economies of China, India, and Russia accounted for much of this total. The demand for iron ore is expected to increase each year into the foreseeable future, with more mining and recycling required to meet it. A relatively small asteroid, approximately 1 km in diameter, might contain 2 billion tons of iron ore—enough to meet the annual global demand for one year. The asteroid 16 Psyche may contain enough iron ore to meet global demands for millions of years. And that is just the iron ore. Spectroscopic analysis of other asteroids shows that they contain nickel, cobalt, copper, platinum, and gold.

Although robots from Earth have flown by or touched down upon a few NEOs, one of the best tools for remotely observing them is the huge radio telescope located in a crater in Arecibo, Puerto Rico. To observe certain characteristics of an NEO, a radar pulse is sent out from this instrument. The reflected return pulse is later received and analyzed.

We now know that there are a few thousand of these objects in the 100-meter size range or larger. They rotate with periods of hours or days. Some NEOs (and main-belt asteroids) are known to have satellites.

There seems to be a large variation in the physical characteristics of NEOs. Some NEOs are very dense; they are most likely composed of heavy metallic elements such as iron and nickel and are similar in composition to Earth's core. Others are stony and of lower density; in composition, at least, they may not be dissimilar to Earth's crust. Some NEOs are very tenuous; they may resemble flying rubble heaps, held loosely together by the aggregate's weak gravitational field. Many NEOs are classified as chrondites; with water, carbon, and various volatile substances, they are most likely extinct comet nuclei.

Now that we have established that NEOs contain essentially limitless raw materials that our civilization so desperately needs, what do we do about it? How do we get access? Most importantly, how do we get the raw materials we need from there to here—affordably? Approaching the problem from an American point of view (since the authors are Americans), we can begin building our in-space mining infrastructure using the next generation of rockets that NASA is developing to support a return to the moon in the late 2010s and early 2020s.

Ares-I is the name given to the rocket that will be used to take American astronauts into space, replacing the aging Space Shuttle after its planned retirement in 2010. Ares-I is designed to ferry only a crew to space, not cargo. The lofted payload will be relatively small—a capsule designed to carry no more than six people. The capsule, now named Orion, will be used to dock with the International Space Station or with a larger vehicle designed to take the crew to the lunar surface. This larger vehicle will be carried into space by the second of the two new rockets NASA is developing, called Ares-V. Ares-V will be a heavy lift rocket capable of lifting 188 metric tons (414,000 pounds) to low-Earth orbit. When working together with the Ares-I crew launch vehicle, Ares-V will be capable of sending nearly 71 metric tons (157,000 pounds) to the moon. For comparison, the Saturn V rocket could send 47,000 kg (100,000 pounds) to the moon. If all goes as planned, both the Ares-I and Ares-V rockets will be coming into use within the next decade. The Ares-V is not yet in production, so many design changes are possible between now and its maiden voyage sometime in the late 2010s. However, its overall configuration is expected to remain basically the same as that shown in Figure 10.3.

Figure 10.3. NASA's Ares-V rocket. (Courtesy of NASA.)

Using a scenario similar to that already being planned by NASA for lunar missions, initial prospecting and mining missions could begin at selected NEOs. It might work something like this: Ares-I launches a crew of six into low Earth orbit. Shortly thereafter the Ares-V launches. Onboard the Ares-V is a docking ring that attaches to the Orion crew capsule lofted by the Ares-I, with the other end designed to attach to the surface of the asteroid. It is called a docking ring instead of a lander because the gravity around a NEO is so small as to be almost nonexistent. Landing on an asteroid will be more like grabbing hold of an outcropping on a climbing wall. It is likely that the docking ring will have to be secured to the surface with cables so that it does not get knocked into nearby space. Also onboard the Ares-V will be an inflatable solar concentrator module that will be securely fastened to the asteroid's surface by the Orion's crew. This module will be responsible for propelling the asteroid into a stable orbit either around Earth or one that permits frequent access as both bodies orbit the sun.

The solar concentrator module can best be thought of as a magnifying glass, focusing sunlight onto the surface of the asteroid, vaporizing the surface materials it encounters, similarly to how children use a magnifying glass and sunlight to burn paper. In this case, the intense heat turns the materials at the burn location into superheated rocket exhaust moving outward to the asteroid's surface, which forces the asteroid to react, or move, in the opposite direction. Thus, we will have built a very primitive rocket. This rocket will operate as long as it remains in sunlight, which can be very often even if the asteroid spins on its axis, and will produce a small but consistent thrust that will change the natural orbit of the asteroid to one that we want. Once the solar-powered rocket is in place and operating, the crew will return to the Orion vehicle and make a safe return trip to Earth.

After the NEO is safely in a high Earth orbit, somewhere outside the orbit of our geostationary communication satellites, the mining can begin. Again, the mighty Ares-V will be used to loft the equipment necessary for mining the raw materials as well as the habitats that will be used by our space miners during their tour of duty "on the rock." The miners will be launched into space on another Ares-I rocket.

Once the raw materials are collected, they must be safely sent back home for processing into the finished products Earth's burgeoning consumers will demand. Since the material will be massive, a highly efficient propulsion system will be required. If we have to send tons of rocket fuel into space to retrieve tons of raw materials, it might not be worth the effort!

One possible way to bring the resources home would be to place them in a container capable of sustaining the high temperatures associated with entering Earth's atmosphere and decelerating from many thousands of miles per hour to nearly a complete stop just before impacting the ground or ocean. The materials required to suitably insulate such a container exist today and are reasonably lightweight. The container may be propelled to Earth in two steps.

The first step would use a conventional chemical rocket to slightly alter the orbit of the container from being alongside the asteroid to one that is slightly lower in its highest altitude (apogee) and much lower in its lowest altitude (perigee). If the perigee can be made to dip below about 2000 km, then a long, thin conducting wire can be deployed to interact with Earth's magnetic field and ionosphere and produce additional thrust. This wire is called an electrodynamic tether. A detailed explanation of how electro-dynamic tethers work and how they produce propulsion is provided in our previous book, *Living Off the Land in Space* (New York: Springer, 2007). For our purposes here, it is sufficient to understand that these tethers produce and carry large currents generated as they move through Earth's magnetic field. It is through this electromagnetic interaction with Earth's magnetic field that the perigee of the container's orbit will continue to be lowered until it dips into Earth's atmosphere, at which time atmospheric drag will take over, causing the final entry, descent, and landing.

Landing in water may be preferable to landing on dry land because of the potential risk of the payload's not reaching the exact landing target. Most of Earth's surface is covered by water, and a near miss in the ocean is not likely to cause any harm. The same cannot be said of landing a multi-ton capsule full of iron ore 100 miles from its intended target, which might or might not be a populated area. Once the payload is recovered, Earth-based industries can do the rest and produce whatever 21st-century products will be manufactured from the ore.

Different strategies are required for different NEO classes. Altering the trajectory of a rubble pile requires different equipment than does moving a metallic NEO in the same fashion. A number of devices exist for NEO trajectory alteration. The solar collector, to be discussed in Chapter 13, works best with extinct comet nuclei or NEOs with extensive dust layers. The mass driver, a type of electromagnetic catapult that ejects NEO material at high speed, would function well with rocky or stony NEOs.

As well as altering the NEO trajectory, several additional dynamical options must be addressed by would-be space miners. Do we de-spin the NEO (reduce its rotation rate)? Do we steer it into high Earth orbit where astronaut teams could deconstruct it in the vicinity of Earth? Or do we instead send portions of the NEO toward Earth, controlled by solar sails, solar-electric engines, or similar devices?

Finally, we might wonder what NEO resources are most profitably mined.

Rock could be used as cosmic ray shielding by interplanetary spacecraft; water ice could be used in the ecospheres of space habitats or electrolyzed as rocket fuel.

It seems most unlikely that the old science-fiction scenario of space prospectors becoming rich after their discovery of a gold- or uranium-rich NEO will ever come to pass. Simply delivering such an object to Earth along a controlled trajectory will greatly reduce the value of the precious, previously rare commodity.

Earth's economy and climate might benefit if NEOs are mined for silicon and other materials from which solar power satellites can be manufactured. Use of such free flying satellites to beam solar power to Earth has certain advantages over moon-based power scenarios.

There are other technologies for asteroid rendezvous, orbit alteration, and material recovery that need to be developed before the first asteroid miners can begin operation. None of the needed technologies appear to be beyond our grasp. What is important to know is that the resources are there, they are accessible with relatively near-term technologies, and can be sent to Earth with little or no environmental impact.

The next order of business is Earth protection. Equipment, possibly resembling that to be discussed in Chapter 13, would be erected to slightly alter the NEO's solar orbit so as to reduce the threat to Earth.

Comets

French Queen Matilda, in or around the year 1077, allegedly created the Bayeux Tapestry (Fig. 10.4). Seeking to honor William the Conqueror and the 1066 Norman invasion and conquest of England, Matilda commissioned the tapestry for placement in the Bayeux Cathedral. This 50-centimeter by 70-meter work of art is rich in historical content, telling the story of the invasion and the events leading up to it.

In reality, and as chronicled on the tapestry, Halley's comet appeared just before the invasion, signaling impending doom for the English. Surely this wispy celestial figure was an omen from God—at least that is the way medieval people interpreted comets and their often majestic appearances in the sky. In our day, we know far more about comets, where they come from, and why they appear as they do in the skies above, though perhaps we, too, should take a cue from the Bayeux Tapestry and understand their existence as an omen—though not of doom, but of plenty.

Like asteroids, comets are often remnants of the early solar system, orbiting the sun for billions of years, often having no direct interaction with any other body in the solar system. Unlike asteroids, they are made from an ice-dust-rock ensemble that produces a visible atmosphere when exposed to

Figure 10.4. A man-at-arms warning Harold of the disastrous omen of Halley's comet. A detail of the Bayeux Tapestry (ca. 1070–80), Bayeux, France.

solar radiation. Comets are often spectacular and have been objects of study since there were humans to observe them. Figure 10.5 shows Comet Neat providing a truly awesome sight in the night sky.

The tail of a comet may extend millions of kilometers into space, always trailing away from the sun due to the pressure of sunlight and the solar wind. By contrast, the nucleus itself is typically only tens of kilometers in diameter. That such a small object can produce a massive and beautiful display is truly one of nature's wonders.

Comets are abundant in the outer solar system and only a few of them have orbits that regularly bring them close enough to the sun for us to observe directly. Most are quietly careening through the outer solar system in nearly circular orbits well beyond the orbit of Pluto. Some are in highly elliptical orbits (meaning that their orbits are shaped like eggs, rather than spherical balls) that allow them to periodically plunge into and then out of the inner solar system, providing us with the sky display described above.

Figure 10.5. Comet Neat shows its splendor in the Arizona sky. (Courtesy of NASA.)

The most famous of these visitors is Halley's comet, which is set to make another appearance in 2061.

What resources will comets ultimately be able to provide for the Earth? Perhaps none—at least directly. But any large-scale industrialization of space will require water. In fact, water may likely be the most important extraterrestrial resource required for our expansion into the solar system. Water is needed to support human life (drinking, food preparation, bathing) and our industrial activities (coolant, cleanser, and raw material). We also need it for fuel in our spaceships. For example, NASA's Space Shuttle's main engines burn liquid oxygen and hydrogen, producing water as the by-product. That big pillar of "smoke" seen at the launch of the shuttle is not smoke at all; it is a pillar of steam. Future spacecraft can be refueled using hydrogen and oxygen obtained from the electrolysis* of water, perhaps made from cometary ice.

---

* *Electroysis* is the process in which an electrical current is passed through water, breaking it into hydrogen and oxygen.

To provide water for use on the moon, we might someday alter the orbit of a comet so that it smashes into the moon. The resulting ice fragments could then supply a future human settlement there for decades or centuries. In a sense, we would be doing on the moon what nature did to Earth so long ago. Earth's abundant water is thought to have arrived here with the impacts of comets into its surface many millions of years ago.

Mars' Moons and Main-Belt Asteroids

Further out in this solar system realm of rock and dust, we come to Mars's tiny satellites Deimos and Phobos and the main-belt asteroids. These small bodies can certainly serve the ultimate needs of a space-based solar-system civilization. But there seems to be no immediate utilization of the immense resources in these objects to help reclaim Earth. So our best strategy might be to study and inventory the resources of these objects for the far future.

The same must also be said for Mars. If we elect to establish colonies on this barely habitable world and gradually transform its environment, Mars will be a net importer of resources for millennia.

In the Realm of the Giants

Beyond the main-belt asteroids, at five times the distance of Earth from the sun, is giant Jupiter. At 318 times Earth's mass, Jupiter is indeed king of the solar system. However, the mass of our sun is about one thousand times the mass of this giant world.

Far from the sun, we are now in the zone of gas and ice. Jupiter and its smaller colleagues Saturn, Uranus, and Neptune are giant gas balls. Jupiter and Saturn are mostly hydrogen and helium. Uranus and Neptune are rich in methane and ammonia. All of these planets have ring systems and are accompanied by many satellites. Many of the satellites of these giant worlds are coated with layers of water ice. Some of these satellites are as large as small planets; others are captured asteroids or comets.

Space miners seeking to tap the resources of the giant worlds also must beware of multiple hazards. For example, Jupiter is equipped with a strong magnetic field and extensive radiation belts.

It seems unlikely that future space miners will venture this far from the sun in search of water ice or silicon for solar photovoltaic cells. But there is one tantalizing resource in the atmospheres of the giants that may someday be tapped for terrestrial use.

If we establish a nuclear-fusion–based economy, a source of helium-3 would be a good thing. The atmospheres of the giant planets contain this isotope in approximate solar abundance—concentrations of about one part in 10,000.

In his book, *Mining the Sky* (Reading, MA: Addison-Wesley, 1996), John Lewis describes conceptual plans to mine the upper layers of the giant worlds for this resource. A system of robotic shuttles would fly the multiyear round trips between Earth and the giant world of interest. Suspended by balloons, helium mines would be dropped into the upper atmospheric layers of the large worlds. On their return to Earth, the shuttles would be loaded with this precious form of helium. In light of the interplanetary capabilities currently under development by many nations, this load is far from enormous. Only one Apollo-massed interplanetary shuttle loaded with helium-3 must reach Earth each month to supply global energy requirements in this manner.

Choices

As we look further into the solar system, we come to the enormous resource pools on the Kuiper Belt dwarf planets and the Oort Cloud comets. But most of this material is best thought of as a resource to help maintain a future solar-system civilization, rather than for import to the Earth.

Our use of space resources will depend on what choices we collectively make as citizens of a global civilization. If we desire a solar-powered world, two choices are lunar and NEO resources. Do we develop the moon because it is close and we have visited it in the past? Or do we instead bypass the moon because NEOs must be moved around to reduce the risk of collisions with Earth—and if we have to move them, why not use them?

Alternatively, a substantial fraction of our planet's future energy requirements may be produced by thermonuclear fusion. In such a case, if breakthroughs do not allow tapping the solar wind directly and if lunar concentrations of helium-3 are too low, robotic exploitation of the giant planets may become a reality.

Unfortunately, we have no idea what direction humanity will take. But it is comforting to know that the resources we will need are available in space and that there are many options to obtain them for Earth's benefit.

Further Reading

For more about the possibility of future wars over natural resources, we recommend *Resource Wars: The New Landscape of Global Conflict*, by Michael T. Klare (New York: Holt, 2002). A good source of data on solar system objects is Katharina Lodders and Bruce Fegley, Jr., *The Planetary Scientist's Companion* (New York: Oxford University Press, 1998). One of the best sources on solar-system resources is the very readable John S. Lewis,

Mining the Sky (Reading, MA: Addison-Wesley, 1996). A somewhat more venerable treatment of the same topic, authored by an ex-astronaut, is Brian O'Leary, *The Fertile Stars* (New York: Everitt House, 1981).

Technical treatments of NEO resources and NEO mining possibilities can be found in two scientific papers by Richard Gertsch, John L. Remo, and Leslie Sour Gertsch, "Near-Earth Resources" and "Mining Near-Earth Resources," published in the proceedings of the United Nations-sponsored conference at which they were presented: John L. Remo ed., Near Earth Objects, *Annals of the New York Academy of Sciences*, 1997, Vol. 822.

radio continuum (2.5 GHz)

Power from the Sun

Busy old fool, unruly Sun,
Why doest thou thus,
Through Windows and through curtains, call on us?

John Dunne, from "The Sun Rising"

386,000,000,000,000,000,000,000,000 watts. By any measure, that is a lot of power, and that is the sun's power output every second of every day (Fig. 11.1). To make this number easier to work with, scientists write it as 3.86×10^{26} watts, or 3.86 followed by 26 zeros. The amount of that energy reaching Earth, which is 93 million miles away from the sun, is 1.74×10^{17} watts, or approximately 1368 watts per square meter. By way of comparison, in 2005 the total power output of the entire human race was approximately 1.5×10^{13} watts! In that year, we generated a mere 0.009 percent of what the sun sends to Earth each second. If we can tap into this tremendous energy source, then surely the global energy problem can be solved.

The truth is, we do use this energy already. As taught in author Les Johnson's daughter's seventh grade science class, all living things on Earth derive their energy from the sun, though we humans do so indirectly. In photosynthesis, Earth's plants convert sunlight to carbohydrates for use in their growth and for their ultimate consumption as food by animals. We are near the top of the food chain, yet all of the foods we eat derived their stored energy from the sun. Do you like steak? The cow from which the steak is cut consumed some sort of grain, which grew using sunlight in the photosynthesis process.

We use the sun's energy in other ways as well. Fossil fuels are formed by the long-term decay of plants and animals that existed on Earth in its ancient past. The energy they collected during their lifetimes is stored in the coal, oil, or natural gas—all of which are called fossil fuels, since they are derived from fossils. When we burn the oil, we are releasing the energy stored chemically in that ancient plant. Think of it as a battery.

The recent trend toward biofuels is yet another example of how we use the sun's energy for power. Biofuels are simply artificial fossil fuels that are

L. Johnson et al., *Paradise Regained*, DOI 10.1007/978-0-387-79986-5_11,
© Praxis Publishing Ltd, 2010

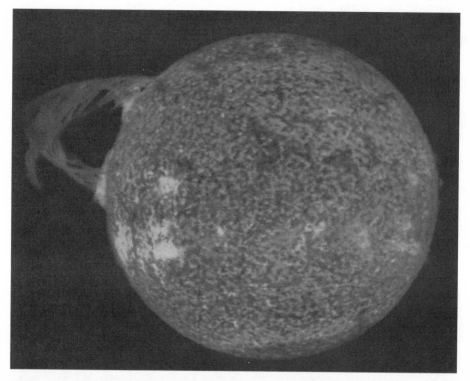

Figure 11.1. The sun emits an enormous amount of energy, enough to meet our energy needs for as long as the sun and Earth exist. (Courtesy of the Naval Research Laboratory.)

manufactured from plant mass. The plants are grown, harvested, and processed to extract fuel that we then burn to run our machines. Unfortunately, the process of manufacturing biofuels involves many different steps, all of which are relatively inefficient. The end product, which is the output of whatever the machine burning the biofuel is intended to accomplish, is produced at the expense of a relatively large energy input.

According to the United States Department of Energy, the average American household uses approximately 14,000 kilowatt-hours (kWh) per year. A kilowatt-hour is the amount of energy a 1-kilowatt appliance would use if left running for 1 hour. For example, a 100-watt light bulb burning for 1 hour would use 0.1 kWh; if it were left on for 10 hours, it would use 1 kWh. While this is currently far more power per person than that consumed by the citizens of the rest of the world, there is reason to believe that rest of the world's per capita energy consumption is rising and will continue to rise in the future as standards of living increase. Burning fossil fuels in automobiles, electrical power plants, or other machines like jet aircraft currently consumes most of this energy.

We can convert sunlight directly into electrical power by using a solar cell. Solar cells use quantum mechanical effects to convert some fraction of the sun's energy falling on them to usable electrical power. Earth-based solar cells can be used effectively for applications that do not require a lot of power, like highway road signs and calculators. But if you need megawatts on a continual basis, solar cells are simply impractical in many locations. Terrestrial solar cells can generate power only during the day, only when the sun is not obscured by clouds and rain, and in locations that do not degrade the materials from which they are made. Unfortunately, there is no place on Earth that gets continuous sunlight and never has bad weather.

What if you could locate these massive solar array farms in a location that is in almost perpetual sunlight with no cloudy or rainy days? You might have a chance at providing continuous power to an energy-hungry population without directly generating pollutants or emitting greenhouse gases. Then you might have a power solution worth considering. Space provides this optimal location to generate electricity for a power-hungry Earth.

In 1973, Peter Glaser proposed using microwaves to beam such space-generated power back to Earth. Others subsequently proposed using lasers or other electromagnetic radiation of various frequencies to beam the energy to Earth for conversion and distribution to the electrical power grid.

What is a microwave? Like sunlight and radio waves, microwaves are simply a part of the electromagnetic spectrum. The only difference between a microwave and a radio wave (from a radio station) is its frequency. In fact, a microwave, a radio wave, and the light emitted by a reading lamp are all essentially the same thing, distinguished only by their different frequencies. All are considered electromagnetic radiation and, as evidenced by the myriad radio transmissions passing through our bodies all the time, many are quite harmless to biological organisms such as humans and most other denizens of the biosphere. And the microwaves used by space solar power satellites can be designed to be equally benign.

Before going into the details of how these systems can be constructed and operated, the overall concept needs to be described.

High above the surface of Earth, very large panels of solar collectors can be deployed to convert sunlight into electricity. There are orbits around Earth, described later, that are in nearly perpetual sunlight. In these orbits, spacecraft can continuously generate electrical power without having to worry about the weather or that pesky darkness that happens every night. The power they generate is then converted to microwaves and beamed back toward Earth. After traveling through the atmosphere, these microwaves are

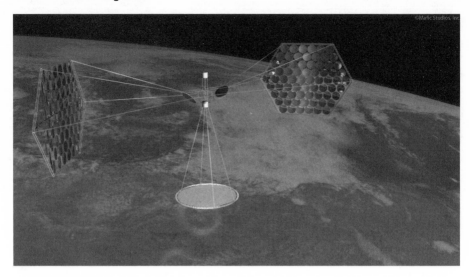

Figure 11.2. Artist conception of an Earth orbiting space solar power satellite beaming power back to Earth. (Artwork courtesy of the National Space Society.)

collected by large antennae and converted back into electricity. Figure 11.2 illustrates what a space-based solar power satellite might look like once it is operational. Generating power in this manner is the ultimate "green" energy. It consumes few resources from Earth, other than those used in its construction. The bulk of the system operates well out of the biosphere, and it does not blight the landscape, as do conventional fossil-fueled power plants. The energy passes harmlessly through the atmosphere to be converted into relatively clean electrical power, which can then be used to power our cities and future-generation electrically powered transportation systems, including our cars and trucks. The power source is the sun and it never sets for our orbiting power stations. Unfortunately, there are still some significant technical hurdles to overcome before this vision can become a reality.

To understand these hurdles, and some options for addressing them, we now take a closer look at each part of a space-based solar power system.

The Orbit

To achieve the goal of nearly perpetual sunlight, a spacecraft cannot be in low Earth orbit. For reference, the International Space Station is located in low Earth orbit at an altitude of about 400 kilometers. It is at about this altitude that all American Space Shuttle flights were conducted. Spacecraft in low Earth orbit circle the globe approximately every 90 minutes, experiencing a complete day/night cycle with each orbit as they do so. Spacecraft at these altitudes pass through Earth's shadow approximately 50 percent of the time and they therefore cannot produce continuous electrical power. There is an altitude, however, at which a spacecraft can orbit Earth and almost never enter its shadow. This particular orbit is called "geostationary" and it is located 22,300 miles above our heads.

A geostationary orbit provides to a spacecraft what its name implies. "Geo," meaning Earth, and "stationary," meaning that it appears not to move as it circles the globe. This seemingly contradictory orbital situation results from the fact that Earth rotates, producing our 24-hour day. In a geostationary orbit, the velocity required for a spacecraft is such that it circles Earth once every 24 hours, matching the orbital rotation rate of Earth. The result is a spacecraft that always remains directly overhead any particular point on Earth's equator. It is sufficiently far from Earth so that it almost never enters Earth's shadow; thus there is no real "nighttime." A spacecraft in geostationary orbit will experience almost continuous daylight.

Imagine a 1-km-wide band encircling Earth's equator in geostationary orbit. The sunlight falling onto this area of space, if collected for just one year, equals approximately the total amount of energy contained in all the remaining known oil reserves in the world today.

Converting Sunlight Into Electricity

There are two good candidates for converting sunlight into electrical power. The first is likely to be most familiar to anyone who owns a solar-powered calculator or other electronic device. In this approach a semiconductor material in an array converts sunlight into electricity directly using quantum mechanical effects. The semiconductor is called a photovoltaic array (PVA), commonly referred to as a "solar array." The leading alternate approach is called solar dynamic power conversion. Solar dynamic systems use the energy in the sunlight to heat a working fluid that then drives some sort of electrical power generator. This approach is very similar to that used in coal-, oil-, and nuclear-fueled power plants on Earth.

Most spacecraft use photovoltaic arrays, converting sunlight into electrical power, and explore the inner solar system using the power they generate. Over time, the efficiency at which these arrays operate has increased, allowing for either smaller solar panels or instruments that require higher power to be used.

The International Space Station (ISS), shown in Figure 11.3, uses multiple

Figure 11.3. The International Space Station on November 5, 2007, with four Solar Array Wings deployed. (Courtesy of NASA.)

solar array wings to produce its own electrical power. Each wing is 34 meters (m) (112 feet) long and 12 m (39 feet) wide. When complete, the ISS will have eight such wings and generate tens of kilowatts of power.

Space-based power conversion efficiencies are likely to be greater than 33% by the middle of the second decade of the 21st century, making the technology very attractive for space solar power applications.

Sending the Power Back Home

Transmitting power through the air without wires is not science fiction. In 1964, pioneer William C. Brown demonstrated on live television that microwaves could power the flight of a miniature helicopter. In 2003, engineers at the NASA Marshall Space Flight Center flew a small, propeller-driven model airplane powered by a laser (instead of microwaves). In 2008, newspapers reported that Japanese researchers began testing a microwave power beaming system as part of their ambitious plans for eventually building space solar power stations.

To make it work in our energy grid, the power generated by the solar arrays onboard the spacecraft will have to be converted to electromagnetic radiation and transmitted to a receiving station on the ground. In order to accomplish this, the frequency of the radiation must not be harmful to life, must not damage the atmosphere as it passes through, and must not be absorbed on its way to the surface by the water, carbon dioxide, and other gases in the atmosphere. In addition, to prevent having extremely large antennae to broadcast and receive the power beam, the wavelength of the radiation must also be as small as possible.

When all of these constraints are considered, the leading candidates for power beaming from geostationary orbit to the ground are microwaves and lasers. Microwaves can pass relatively easily through our water-laden atmosphere, they can be of a wavelength that is not absorbed by the human body, and they can be generated and received by antennae that are of sizes that can at least be mentally conceived.

A laser optimized for power beaming would have greater difficulty with attenuation as it traverses the atmosphere but it would require less of an infrastructure on the ground. For example, the ground-based receiver would only be a few tens of meters in diameter versus a microwave rectenna with diameter of 1 to 10 km. If a method to decrease atmospheric losses is found and implemented, then laser power transmission would become the technology of choice.

Antennas for Transmitting and Receiving

As described above, space-based solar power stations will be sending the power back to Earth using microwaves. Recall that a microwave beam differs from a radio wave only by its frequency. They are fundamentally the same, except for their rate of oscillation (frequency) and wavelength, and except that radios need an antenna to transmit and a separate antenna to receive. In this case, the space-based transmitting antenna will need to be 1 to 2 km in diameter and the receiving antenna on the ground about ten times larger.

A note about building large structures in space is warranted. It is far easier to build large, even massive structures in space than it is to build similar structures on the ground. In space we are free from the unrelenting tug of Earth's gravity and can make our structures out of extremely low mass materials that would not even hold up under their own weight if they were used on the ground. Without wind and rain, we do not have to worry about many of the same size-limiting design constraints that we do in terrestrial applications. Prototype large structures have already been built in space.

The ISS became operational in the year 2000 and, when it is completed, it will be approximately the same size as a football field. In the lobby of a building at NASA's Marshall Space Flight Center in Huntsville, Alabama, is a scale model of a football field (University of Alabama versus Auburn University, of course), within which sits a same-scale model of the ISS. This allows casual visitors to know the scope of this grand engineering endeavor.

In addition to the space station, we humans have built and flown structures 20 km in length. In 1996, the Small Expendable Deployer System (SEDS) unfurled a 20-km-long, 0.075-cm-diameter cable from a Delta II rocket. This feat was duplicated several times by NASA, and comparable feats were demonstrated by other space-faring nations. This is certainly not a 20-km-square solar array farm, but it is an essential first step toward the capability of building one in the future.

Distributing the Power

By comparison, this part of the system is simple. The power collected by the ground-based antenna must be regulated (voltage, frequency, and overall quality) for connection to the power grid. Highly efficient power conversion and regulation technologies already exist and are in use throughout the world. Adapting them to this application should not be difficult.

The Catch

Yes, there is always a catch. This virtually limitless, completely renewable, continuous power system will be expensive to develop and launch into space. The launch requirements alone, at today's prices, are astronomical (pun intended). A 4-gigawatt (GW) (4-billion-watt) power station would weigh in excess of 2 million kilograms and require perhaps twenty launches of NASA'S planned Ares-II rocket just to get the construction materials into space. Then it would have to be assembled, probably requiring humans since no matter how capable our robots may be, there is no substitute for having a person at the site in case something goes wrong. At today's prices, the launch costs alone could exceed $20 billion. (But with trillions of dollars being used in 2009 to bail out the world's financial institutions, this seems like a worthwhile and affordable investment!)

Then there are the rest of the infrastructure costs. How much will it cost to build that spacecraft and solar arrays, the antennae, and the ground support equipment? These costs could easily total in the billions of dollars. For lack of detailed accounting analysis, let's say this infrastructure cost is on the order of half the launch cost, placing the total system price at approximately $30 billion. For comparison, using today's dollars, a coal-fired power plant would cost "only" hundreds of millions of dollars.

Space-based solar power is clearly not cost competitive—yet. Improvements in solar array efficiency seem to be occurring on a regular basis. As of this writing, inventors are claiming efficiencies greater than 35 percent. Previous studies of space solar power assumed much lower efficiency solar cells, therefore requiring many more cells to produce the same amount of power as higher-efficiency ones. With these cells, the amount of mass to be carried to space will decline, resulting in a decline in the launch cost. But it would have to decline dramatically to make a significant difference in the estimated multibillion-dollar cost.

What can possibly make this affordable? Well, that all depends on the cost of energy and how much of a value we place on the environment. The cost of energy production is not as simple as dollars, euros, or yen. What is the cost to the planet of the strip mining required for the coal we burn in our thousands of power plants? What is the payoff in reduced defense spending that will result from us not having to depend on the volatile Middle East for oil to generate electrical power? How much is it worth to eliminate the acid rain associated with the burning of fossil fuels? What benefits will we reap from a power system that produces no greenhouse gases? The authors contend that when the real societal costs are considered, as well as the real monetary cost from end to end, space-based solar power begins to look like a winner. It is an expensive winner, but a good investment nonetheless.

On-Demand Power: A Niche Application

In addition to supplementing the power grid, space-based solar power can be used to send power to localized areas that have an urgent need for electricity. It is perhaps for these applications that space solar power will find its first use due to the reduced overall power demand and the commensurate lower-mass systems that will be required.

Following a natural disaster, one of the first priorities is getting power into the affected area. While utility crews busy themselves in restoring power from the grid in a process that can take from weeks to months, urgent care providers have an immediate need for electricity from the moment they arrive. Lives are often at stake. What if a constellation of relatively small orbital power satellites are placed in low to medium Earth orbit—much like the satellites in the Global Positioning System—so that one or more of them have line-of-sight power transmission capability to just about any point on

Earth? Power generated by these satellites could be beamed to the affected areas as soon as a ground station receiver is put into place. This would reduce the long logistical convoys currently required to keep fuel flowing to the first responders, making them independent of the rest of the world for immediate power.

Further Reading

The National Space Society has an excellent online space solar power Web site: http://www.nss.org/settlement/ssp/. Another excellent resource is "Space-Based Solar Power as an Opportunity for Strategic Security," report to the National Security Space Office, October 10, 2007. We also recommend Peter E. Glaser, Frank P. Davidson, and Katinka Csigi, *Solar Power Satellites: A Space Energy System for Earth* (New York, Wiley, 1998).

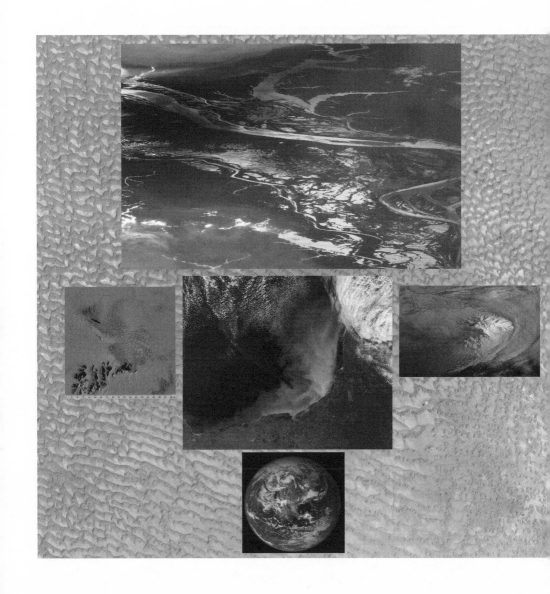

Environmental Monitoring from Space

12

The people know the salt of the sea
And the strength of the winds
Lashing the corners of the earth.
The people take the earth
As a tomb of rest and a cradle of hope.
Who else speaks for the Family of Man?
They are in tune and step
With constellations of universal law.

Carl Sandburg, from ''The People Will Live On''

In the late 1990s, the author Frank White visited Huntsville, Alabama, and lectured at the local university on what he called the "overview effect." White had just authored a book by that title and was touring the country speaking on the subject and promoting his book. According to White, most of the world's astronauts experience an epiphany when they first gaze back from space upon this wonderful cradle of life we call Earth. With its magnificent blue oceans, swirling cloud formations, and obvious land formations, these space explorers experience a sense of belonging to the globe, not necessarily as citizens of the United States, Russia, or whatever country was responsible for their making the journey, but as human citizens of planet Earth. From space, without the human-created boundaries dividing the landforms, there is no obvious distinction between the United States and Mexico, between Israel and Syria, or between any other bordering states. Those who have experienced the overview effect claim to be forever altered in their perspective regarding not only countries, but also their peoples and the planet.

Many who have not been to space can experience this vicariously by looking at the famous "Earth Rise" photograph taken by the Apollo 8 astronauts during their journey to the Moon. Shown in Figure 12.1, this image is credited by many as awakening their awareness of how apparently fragile this lonely jewel of life may be among an uncaring and deadly cosmos. Other than an ethereal and somewhat elusive moment of spiritual

L. Johnson et al., *Paradise Regained*, DOI 10.1007/978-0-387-79986-5_12,
© Praxis Publishing Ltd, 2010

Figure 12.1. Earth, as seen by the crew of Apollo 8. (Courtesy of NASA.)

awakening and often-needed perspective broadening, of what use can this and other photographs be for monitoring and protecting the planet? A large scientific community that is engaged in the field of remote sensing provides the answer to this question.

Remote sensing is the term used to describe virtually any method of viewing something from a distance, as opposed to being physically present wherever the measurement is being made. Satellite remote sensing describes the functional ability to detect electromagnetic radiation from Earth's surface or atmosphere and to understand what the observed radiation's characteristics mean regarding the conditions from wherever they originate.

This chapter describes how our ability to access space helps us understand, predict, and prepare for local and global environmental change using various remote sensing techniques and technologies.

Figure 12.2. These pictures, taken in 1973 and 2003, respectively, show how civilization has consumed forestland in Paraguay and Brazil. The Iguazu National Park in Argentina, which is protected from development, remained intact. (Courtesy of the United Nations Environmental Program.)

Visual Images

Since we humans first ventured into space with either our robotic emissaries or in person, we have been taking pictures. In fact, we have taken lots and lots of pictures. As a NASA employee, author Les Johnson is fortunate to engage in regular mission debriefings with astronauts when they return from space, having been aboard the Space Shuttle or the International Space Station. Their debriefings are always in the form of a slide show, reminiscent of one created to describe a family's recent trip to the beach, showing all aspects of the mission from preflight preparations to touch down. A large number of their photographs are of Earth from space. Some of the photos are taken with nothing more than a high-grade commercially available camera; others are taken with high-resolution cameras designed for that very purpose. Whatever their source, they show the various landforms on Earth, and often pay particular attention to whatever geographic region the astronaut photographer calls home. In addition, over the half-century of space flight, these pictures tell an evolving environmental story.

The United Nations recently released a compilation of such pictures in a book titled, *One Planet Many People: Atlas of Our Changing Environment* (New York: United Nations Environment Program, 2005). One of the many striking images from the compilation shows what happened to unprotected land near the Iguazu National Park in Argentina between 1973 and 2003 (Fig. 12.2).

Spring Comes Earlier Each Year

According to the April 1, 2008, issue of *Science Daily,* spring is arriving about 5 days earlier to Eurasian forests than it used to. Using satellite data, researchers were able to track the appearance of leaves as a function of time over several spring seasons, showing an unmistakable trend toward spring arriving progressively earlier each year. After examining two decades of data, scientists learned that not only is spring arriving earlier than it used to, but also that fall is becoming tardy, arriving 10 days later than it did just two decades ago. A similar trend is seen in data collected over North America.

Will an earlier spring and later fall result in a more productive growing season for farmers? Or will it usher in a longer dry season with more extreme temperatures? The verdict is still out, but one thing is for sure: with satellite monitoring, we will be able to understand what is happening as never before.

Biodiversity and Transgenic Crop Monitoring

Genetic engineering and space technology at first thought might not be considered to have much in common, other than owing their existence as fields of research to the scientific revolution. But it is possible that one may enable careful monitoring of the other to ensure compliance with environmental regulations and to project future food crop health and availability.

In recent years, taking advantage of the biotechnology revolution, agriculture companies are developing food crops that are genetically modified to resist certain pests. One of the first commercial successes in this area is an insect-resistant strain of corn called "Bt corn," which is genetically engineered corn containing bacterial genes that express an insecticidal protein from *Bacillus thuringiensis*. In theory, use of Bt corn will increase crop yield due to its not being consumed as readily by insects. Proponents also believe Bt corn will benefit the environment because fewer pesticides will be required; if the food crop is inherently bug resistant, why would a farmer need to use as much pesticide?

Rather than allow wholesale adoption of this new strain of corn, farmers are planting Bt as only a portion of their overall crop. They are also, for the most part, isolating it from their regular corn crop. From the farmer's point of view, this makes good sense. After all, it is a new product, and what if it does not work as advertised? From an environmentally conscious consumer's point of view, it is also good to understand how this corn might crossbreed with regular corn and, perhaps, spread beyond the confines of where it was initially planted.

The U.S. Environmental Protection Agency (EPA) would like to monitor these crops for a variety of reasons and to answer some very pertinent questions:

- Is the Bt corn actually as pest resistant as advertised?
- Will Bt corn spread or crossbreed with regular corn? Alternatively, can it be geographically controlled?
- Are farmers planting this new corn as planned? Or are some cheating and planting more than stated?

There are not enough field agents to monitor the 25 million acres of corn planted across Middle America. What is needed is a way to monitor large areas remotely, preferably from the air or space.

The EPA is working with NASA to determine if the light reflected from genetically modified Bt corn is sufficiently different so as to allow it to be observed from space using a technique called hyperspectral imaging. When the human eye or a scientific instrument observes an object, the observation is actually of the light reflected from the object that reaches the eye or the instrument. In the case of most vegetation, we observe green light. While the human eye can distinguish various shades of green, our scientific instruments can detect many more reflected colors. Each different shade or color corresponds to a different wavelength of light. When these multiple colors are examined simultaneously, they are considered hyperspectral. In the case of the observations of interest to EPA and NASA, approximately 120 colors are observed simultaneously.

If this as-yet-unproven technology is viable, it will provide a large-scale monitoring capability to inform the grower and governmental regulatory agencies of potential pest resistance development in the Bt corn, its ability to be geographically contained, and exactly which crops entering the food supply are transgenic. Flying regularly overhead, hyperspectral instruments aboard spacecraft may provide the data needed. In addition, this capability will provide decision makers with vital data on the annual health of food crops as climate shifts and both crops and pests adapt to their changing environmental niches.

Global Sea Levels

Approximately 70 percent of Earth's surface is covered by water. Another statistic that is not as well known, but is very pertinent as we face the very real possibility that global sea levels will rise significantly in the next century, is that over half of the world's population lives within 100 kilometers of an ocean. If sea levels do rise, then a large percentage of the total population will be affected.

In the recent geologic past, sea levels have fluctuated a great deal. During the last ice age, which occurred about 20,000 years ago, sea levels were about 120 meters lower than today. This happened because an enormous amount of water was contained in massive ice sheets that covered parts of North America and northern Europe. By contrast, during the previous interglacial period, about 125,000 years ago, sea level was at least 4 meters higher than today.

Global warming can produce a rise in sea level through the thermal expansion of seawater and the net melting of glacial ice. We are careful to use the word *net* when describing the melting of sea ice. What counts is how much more ice melts than forms in any given time period (see below).

Ground measurement of sea level changes due to environmental effects is not straightforward. The following events can affect any given sea level measurement:

- The tides (cyclical and predictable)
- Atmospheric pressure (how hard the atmosphere pushes down upon the water beneath it)
- Winds (particularly during storms)
- Changes in ocean currents (where large volumes of water flow at any given time)
- Seasonal changes in density (due to temperature changes and the amount of salt dissolved in the water)
- The relative height of the measuring device (is the land sinking due to erosion?)

The TOPEX/Poseidon and Jason 1 spacecraft have been using radar to monitor sea levels since 1992. Satellite data, combined with that from tide gauges (Fig. 12.3) shows that Earth's mean sea level has increased almost 15 cm over the last 50 years. When looking at the graph, it is important to note

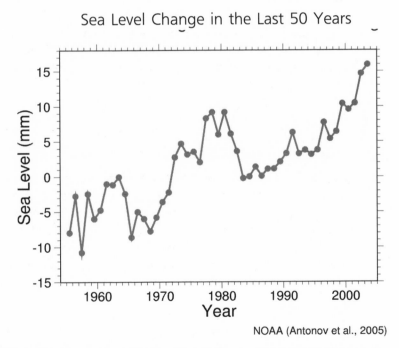

NOAA (Antonov et al., 2005)

Figure 12.3. The graph shows the sea level rise as observed from space. (Courtesy of NASA.)

that the satellite data are different from what was recorded in ground-based tide gauges. The possible causes for the discrepancy range from geologic changes in gauge height to satellite instrument calibration errors. Regardless of the validity of the absolute value of the changes, the trend line is what is of most interest, and that trend is toward increasing ocean sea levels.

Global Ice Measurements

Much of the world's fresh water is locked in glacial ice at either the South Pole or in the Greenland Ice Sheet. If the net amount of ice in these glaciers changes, so will the level of the world's oceans. Satellite observations of the Greenland Ice Sheet show an increase in the so-called melt zone along its edge. This region is still mostly ice, but an increasing portion of its surface area is melting during the summer months, sending rivers of water flowing into the sea. Figure 12.4 shows numerous melt ponds and rivers dotting the surface of the Greenland Ice Sheet.

Satellite observations also show that the thickness of the ice in Antarctica is increasing, taking water from the sea and locking it away in increasingly thick ice near the South Pole. An open question is whether or not the rate at which the ice is getting thicker at the South Pole will balance the amount melting in Greenland and in the Arctic. The net difference will help us understand the potential contribution to the recorded sea level rise from changing environmental conditions at Earth's poles.

Atmospheric Temperature Monitoring from Space

Spacecraft are now used to monitor the temperature of the atmosphere at various altitudes. They do not directly measure the temperature, as they must fly above the atmosphere in the vacuum of space. Instead they observe changes in other atmospheric parameters and infer (calculate) the temperatures at various altitudes. The atmospheric temperature profiles obtained depend on the technique used to make the calculation, and there are different technical approaches that often produce somewhat different results.

In fact, there is an apparent discrepancy between what the satellites see and what thermometers on the ground measure. Analysis of global atmospheric temperatures from space shows no significant warming in the last 30 years. But when the data from ground stations around the world

Melt Pond

Figure 12.4. This image, taken by the Landsat 7 satellite, shows numerous melt ponds and surface streams on the Greenland Ice Sheet. Prior to the mid-1990s, these ponds and streams were rarely observed. (Courtesy of NASA.)

are consulted, a different story emerges—one showing a steady increase in near-surface atmospheric temperatures. At the time of this writing, there is no consensus regarding the cause of this data disagreement. More data are needed from both sources—and some new theories.

Desertification

The transformation of once-vegetative or agricultural land into desert is called desertification. Often directly brought about by human activity, and significantly influenced by climate change, large portions of once-arable land and inland lakes are being transformed into deserts. Activities attributed to cause this burgeoning loss of useful land include urban development, overcultivation, poor irrigation practices, overgrazing, and overt destruction of vegetation. Examples of overt destruction would include clear cutting of forests without regard to reforestation or burning of forests for land cultivation, often resulting in only a few successful growing seasons on the recently cleared land before it is subsequently depleted.

Whenever productive or potentially productive agricultural land becomes desert and unusable, a local scarcity of food is likely to develop. In some locations, particularly in developing countries with limited resources and mobility, the results can be catastrophic. In the 1970s and 1980s, the Sahara Desert grew in size, displaced local and formerly self-sufficient populations and turned them into refugees.

In the early 1970s, the United States launched a series of Landsat satellites which began inventorying land use across the globe. By looking at the data from Landsat and other, more recent, satellites, it is possible to piece together a picture of how much land has been lost to desertification over time. The left side of Figure 12.5 shows an image of Chad Lake in Africa taken from space in 1972. The right side of the figure is an image of the same portion of Chad Lake taken in 2001. When you compare the pictures, you can see that a once water-filled lake has been turned into a marsh, overgrown with vegetation. If the water levels do not soon begin to rise, and the water loss trend continues, this lake will soon be dry. And what of the farmers and human population nearby that depend on the lake water? They will potentially become a displaced, migratory population.

Lest one think that desertification is a problem only for the developing world or just in Africa, as a resident of the southeastern United States, author Les Johnson has to look only a few hundred kilometers to the east to see a similar problem occurring in Georgia, near the growing city of Atlanta. Atlanta uses water from Lake Lanier to supply much of its needs. The

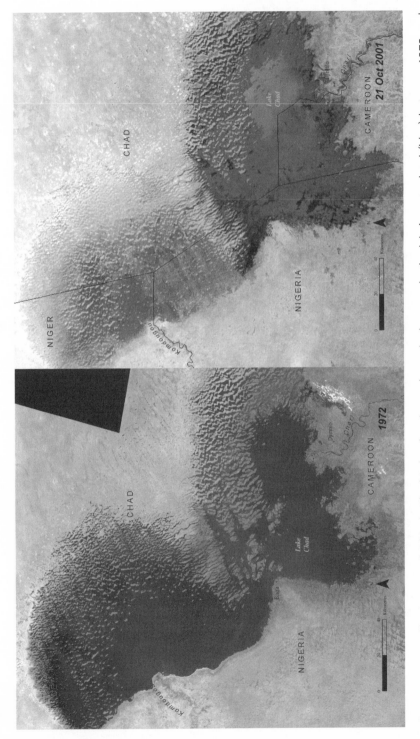

Figure 12.5. Satellite images of Africa's Chad Lake showing a dramatic decrease in the lake's water level (blue) between 1972 and 2001. (Courtesy of the United Nations Environmental Program.)

problem is that Lake Lanier is running dry due to the severe drought that has gripped the region for several years. With a major urban population pulling water from the lake and nature not providing sufficient rainfall to replenish it, the lake's water level is dropping dramatically. At this writing, there is disagreement among the states of Tennessee, Alabama, and Georgia over who owns the water in nearby rivers and how that water will be allocated. The debate has been acrimonious—and this is within the borders of a prosperous and stable country. Imagine the issues to be resolved should these have been separate countries!

Conclusion

Monitoring the global environment can best be accomplished by using the unique vantage point of space for remote sensing in addition to comprehensive ground and aircraft observations. When taken together, the data will provide scientists with a better understanding of our planet and its changing climate and environmental conditions. Continued observations will also allow us to understand how much progress we are making as we alter the way humans interact with Earth and march toward a sustainable future in which the direction of environmental change is not toward degradation but toward regeneration and renewal.

Further Reading

Frank White's book, *The Overview Effect,* is a mind-expanding experience that is highly recommended reading. For more information about the U.S. Environmental Protection Agency's interest in monitoring the growth and spread of transgenic crops, see an excellent summary article in the September 11, 2003, issue of *Nature* titled, "US Reflects on Flying Eye for Transgenic Crops." A sad but educational "must read" is the United Nations publication, *One Planet Many People: Atlas of Our Changing Environment* (New York: United Nations Environment Program, 2005).

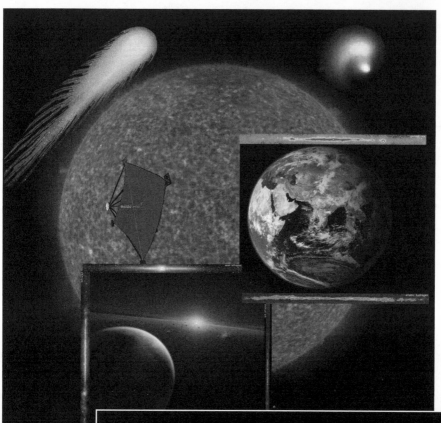

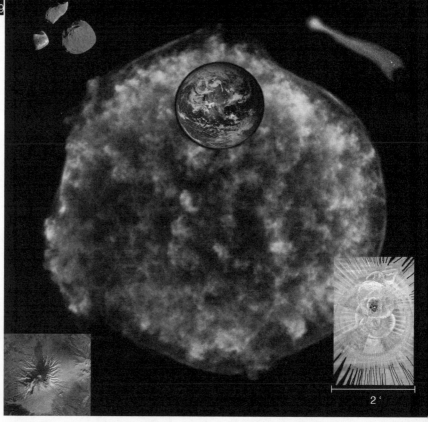

2

Protecting Earth

Full fathom five thy father lies:
Of his bones are coral made;
Those are pearls that were his eyes;
Nothing of him that doth fade
But doth suffer a sea change
Into something rich and strange.

William Shakespeare, from *The Tempest*

Intense tempests of celestial origin have blown through the skies of Earth, obliterating landscapes and sending towering tsunamis through the oceans. Fallout from such events has extinguished vast numbers of living organisms; some of these have been altered by geological processes into fossils.

For the first time, it is almost in our power to prevent or alleviate these events. But the action of protecting Earth from certain types of cosmic catastrophes will change us into something "rich and strange": a spacefaring species.

Catastrophes: Terrestrial and Celestial

There are many types of catastrophes that have threatened Earth life, or may threaten Earth life in the future. Some are of terrestrial origin, and some are of celestial origin. Only some can we alleviate with our advanced technologies.

One terrestrial catastrophic event that may have resulted in mass extinctions in life's long history is the super-volcano. It is not impossible that tectonic forces have on rare occasions produced a volcanic eruption dwarfing Vesuvius and Krakatoa as the atomic bomb dwarfs a hand grenade. If something like that occurred today, civilization (and perhaps humanity) would be doomed by the long-duration obscuration of sunlight that would put Earth in a freezer. We cannot predict such an event and we have no way

L. Johnson et al., *Paradise Regained*, DOI 10.1007/978-0-387-79986-5_13,
© Praxis Publishing Ltd, 2010

of preventing it; perhaps a few representatives of humanity and other species might survive in underground shelters or aboard space habitats.

But super-volcanoes, as destructive as they are, are a purely local phenomenon. There is no way that a terrestrial volcano can affect hypothetical biospheres elsewhere in the solar system, or on planets circling other stars.

Stellar explosions are one class of nonlocal cosmic catastrophes. When a star considerably more massive than the sun reaches the end of its hydrogen-burning life, stellar fires dim and the star collapses upon itself. But the collapse increases temperature and density in the star's interior, which allows a host of energy-releasing thermonuclear reactions involving elements more massive than hydrogen and helium.

In a cosmic instant, the dying star becomes a supernova, converting perhaps 1 percent of its mass into energy and outshining most of the stars in its galaxy for a few days or weeks. A significant fraction of the energy emitted by a supernova is in the form of gamma rays and x-rays. In the case of solar emissions, Earth's upper atmosphere shields us from harmful radiation. But the intense blast of radiation from a nearby supernova might overwhelm this atmospheric shield, extinguishing most life on our planet's surface. A nearby supernova explosion would produce very bad effects. Once again, if we had sufficient warning, our best option would be to dig a deep hole.

It is possible that nearby supernovae caused some of the mass extinctions in the fossil record. Perhaps subterranean life forms then had the opportunity to recolonize our planet's surface.

Happily, most massive stars approaching the supernova stage are very distant from our solar system. One candidate for a supernova progenitor in the near future with a mass of 100 suns is Eta Carinae. Fortunately, this unstable giant resides at a comfortable distance of 7500 light years (one light year is approximately 63,000 times greater than the average Earth–sun separation, or 63,000 astronomical units).

A typical supernova might be bad for life on planets within a few hundred light years. But there are rare cosmic events—so-called gamma ray bursts—that are even more powerful. They have been observed in other galaxies using gamma-ray telescopes in Earth orbit, and may be due to the collapse of super-massive stellar cores into spinning black holes. Happily, there is no evidence to date that such monsters lurk dangerously close to our solar system.

There are no technological insurance polices to protect us from galactic catastrophes such as nearby supernovae or gamma ray bursts. But we have reached the point at which we might be able to do something about the danger posed by asteroid and comet impacts.

Menace from the Skies

Most icy comets currently reside in the distant, frigid Oort Cloud,* spending most of their existence tens of thousands of astronomical units from the sun. These remnants of the solar system's formation would spend eternity in the far reaches of space if it were not for the fact that stars, including our sun, revolve around the center of the galaxy. Our sun requires about 250 million years to complete one circuit through the Milky Way galaxy.

Once every 100,000 years or so, a sun-like star makes a random close approach to our solar system, passing through the outer fringes of the Oort Cloud. On these occasions, the solar orbits of some comets will be altered. A fraction of the affected comets will be flung out of the solar system; others will be directed toward the inner solar system, perturbed into highly elliptical paths requiring tens of thousands of years to complete one orbit around the sun.

As one enters the inner solar system, solar heating causes gas and dust to evaporate from its 10- to 20-km nucleus. A spherical coma consisting of evaporated water, ammonia, and methane gas, with dimensions of perhaps 10,000 km, surrounds the tiny nucleus. Solar radiation pressure pushes one or more tails from the coma, each perhaps 100 million kilometers in length; comet tails always are directed away from the sun.

If one of these comets targets Earth, it will approach from deep space or near-solar space at around 40 kilometers *per second*. Warning time of the impending collision will be quite limited.

Space is big, and Earth impacts by long-period comets are not frequent. But there are other sky objects to be concerned about. Another source of potentially dangerous icy objects is the Kuiper Belt, a region of icy, comet-like dwarf planets including Pluto, extending from about 30 to 50 astronomical units from the sun. The larger of the Kuiper Belt objects (KBOs) are in relatively stable orbits, never entering the inner solar system. But collisions and giant-planet alignments sometimes produce fragments or alter orbits. Then, small KBOs can enter the inner solar system as short-period comets, comets with orbital periods measured in years or decades. A few of these have perihelia (closest approaches to the sun) within Earth's orbit. Because of their varying orbital inclinations, meaning that their orbits do not lie within the same plane as Earth's orbit, most known KBOs will never threaten our home planet.

---

* The Oort Cloud is a spherical cloud of comets believed to orbit the Sun at a distance of about 50,000 Astronomical Units.

Figure 13.1. Asteroid Ida appears to be one large piece of rock. (Image courtesy of NASA.)

Rocky and stony space objects, the asteroids, are generally located closer to the sun than are the comets and KBOs. Most of them reside in the asteroid belt, which extends from about 2 to 5 astronomical units from the sun. The smallest of the asteroids are about the size of boulders; the largest approximate the size of Texas. To make matters more complicated, there are at least two types of asteroids: those that appear to be solid rock (Fig. 13.1), and those that are essentially "rubble piles" held loosely together by their weak gravitational attraction (Fig. 13.2).

But collisions and giant-planet perturbations have altered the orbits of some of these. Several thousand near-Earth objects (NEOs) between about 20 meters and 40 kilometers in size pass close by Earth. Some of these have the spectral signatures of rocky and stony asteroids; others are most likely extinct comets.

Although the threat of Earth-impacting comets cannot be completely ignored, it is the NEO population that represents the greatest threat. We are beginning to observe them systematically and will soon have the capability to deflect Earth-threatening NEOs.

Defending Earth from NEO Collisions

It may have been a 10-km NEO that contributed to the demise of the dinosaurs 65 million years ago. Smaller NEOs in the 100-meter size range

Figure 13.2. A close-up photograph of asteroid Itokawa, which appears to be made of many smaller pieces of rock. (Courtesy of the Japanese Aerospace Exploration Agency.)

pack a smaller wallop, but one of them could still destroy a city. And there are a lot more of the smaller ones.

Potential city-killing NEOs strike Earth at intervals of about a century. The last known impact was in 1908; fortunately, it occurred in Tunguska, a sparsely populated region of Russia.

According to NEO observers, an object dubbed Apophis is scheduled to make a very close pass of Earth in 2029, perhaps approaching within 35,000 kilometers. Although the chances of this few-hundred-meter-wide object colliding with Earth in 2029 are minimal, it will be back in 2036. If Apophis is an extinct comet, gravitational tides caused by Earth may produce a tail during the close approach. And if the reaction to the hot gases emitted by the tail is just right (perhaps we should say "just wrong"), Apophis's solar orbit could be very slightly altered so as to put it on a collision course with Earth on its return seven years later. If the impact of a NEO of Apophis's size occurred on land, a state or small county would be obliterated. If it occurred at sea, the resulting tsunami would drown millions.

There are advantages and disadvantages of the many techniques proposed to divert Earth-threatening NEOs. It is assumed, for this discussion, that the capability will exist to deliver spacecraft in the 100,000-kg mass range, either human-crewed or robotic, to the vicinity of the offending NEO. Many national or international space programs, including those of the United States, Europe, Russia, Japan, China, and India, are developing large interplanetary spacecraft to become operational starting in 2015 to 2020. From the point of view of Earth protection from threatening NEOs, this development is none too soon.

If an Earth-threatening object is from the Oort Cloud, collision warning time will be minimal. In such a case, terrestrial space agencies would probably decide to take the most dramatic action. Earth-launched spacecraft would most likely be robotic, and their payloads would probably consist of what are euphemistically referred to as nuclear or thermonuclear devices.

These 100- or 1000-megaton explosives would be ignited as close to the approaching object as possible, in the hopes of altering the object's trajectory. But there are problems with this nuclear option.

One problem is geopolitical. A Saturn-V class booster capable of delivering nuclear weapons to the approaching comet might be viewed by some nations as the world's most capable ballistic missile. If it were armed with many one-megaton H-bombs instead of one huge explosive, it could serve as a weapon capable of destroying several major cities with a single launch. So planning for the venture would have to be both transparent and international.

But there is a more basic scientific issue with the nuclear option, at least for some comets: it simply might not work.

One class of comets in highly elliptical orbits, the so-called sungrazers, kiss the sun's visible surface during their very close perihelion passes. Some of these have been observed to "calve," or fragment at perihelion.

Another example of a comet breaking into many fragments is Shoemaker-Levy 9. In July 1992, this comet approached Jupiter within 100,000 kilometers. Shoemaker-Levy 9's trajectory was altered by the giant planet's gravity field so that the comet returned to Jupiter on a collision course 2 years later. Astronomers observing the collision of the comet with Jupiter were fascinated to observe that prior to the collision, the comet fragmented into many smaller objects. The likely cause of this fragmentation was the added pull of Jupiter's very strong gravity acting nonuniformly on the comet.

The nuclear option should be applied only as a last resort. Instead of deflecting or destroying an approaching comet, a large nuclear bomb may well cause it to break into many radioactive fragments, each of them still targeting our planet. The end effect would be akin to trading death by a bullet fired from a handgun for death by a close-range shotgun blast. In both cases you are quite dead.

A somewhat less dramatic approach is kinetic deflection. Earth orbits the sun at 30 kilometers per second. NEOs in approximately circular orbits will have about the same solar orbital velocity.

Let's say we launch a robotic spacecraft equipped with a low-thrust propulsion system, such as an ion drive or solar sail. If the orbit of the spacecraft is gradually changed so that it is traveling in the opposite direction to Earth and most NEOs, its relative velocity to an Earth-threatening NEO will be around 60 kilometers per second.

Timing and guidance must be very precise. But if the spacecraft is directed at this relative velocity to impact the NEO, it will pack quite a wallop. According to some model calculations, the momentum change delivered to the NEO might alter the trajectory enough to prevent a collision with Earth, if the warning time is sufficient. Of course, some classes of NEOs may fragment rather than altering course, which would not be a good thing!

Solar Options and the Gravity Tractor

If collision warning time is measured in years or decades, and if NEO mass, trajectory, and other characteristics are sufficiently well known, there are several less dramatic options. These use forces of nature—solar radiation pressure and gravity—to deflect an offending NEO without explosives or impacts.

One of these benign approaches is to encircle Earth-threatening NEO with a reflective solar "parasol." To visualize this approach, imagine that a NEO is like a potato. Astronauts land on the "potato" and stick a number of toothpick-like structures into it. These toothpicks support a spherical, highly reflective, thin-film solar sail, which can be modeled by a sheet of aluminum foil. This increases the model NEO's reflectivity to sunlight and its surface area. The sunlight falling on the reflectors would exert a very small but constant force on the NEO. In the real world, the effects of these changes would be to render the NEO's solar orbit more elliptical. If the warning time is measured in decades, such an orbital alteration might be enough to change a direct hit on Earth into a near miss.

Another solar approach is the "solar collector," which is shown schematically in Figure 13.3. The solar collector is essentially a two-sail solar sail. The collector sail faces into the sun and focuses sunlight on the smaller thruster sail. This thruster directs a concentrated sunbeam on the

Figure 13.3. The solar collector in operation near an NEO.

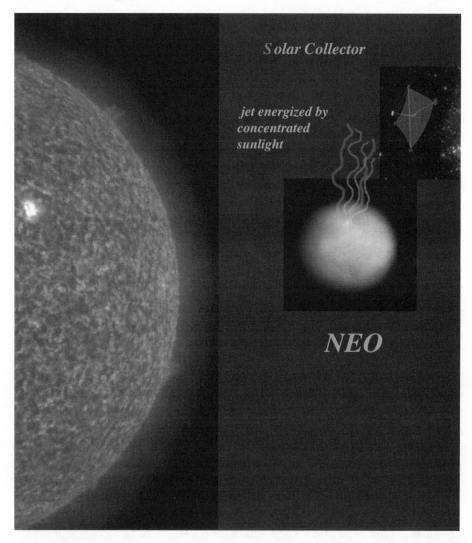

Solar Collector

jet energized by concentrated sunlight

NEO

NEO's surface. If the NEO is coated with layers of dust, soil, or ice, a jet of superheated material (like a comet's tail) may be raised in the direction of the thruster sail. The reaction force to this jet pushes the NEO in the opposite direction (downward in the figure).

For 100-m NEOs and solar collectors in the 50- to 100-m size range, the solar collector seems capable of deflecting NEOs to prevent Earth collisions if warning times are a year or more. But there are issues with maintaining the solar collector on station near the NEO for long time periods and protecting the thruster and collector surfaces from damage by the escaping jet. The solar collector is the only NEO deflection option that allows for the possibility of steering a NEO into high Earth orbit where it could be mined for its resources. For more about asteroid mining, see Chapter 10.

All of the deflection options discussed so far have the disadvantage that they can be applied to only certain types of NEOs. None of them would be effective in the case of a rubble-pile NEO, for example. A very interesting concept that would work with any type of NEO is the gravity tractor.

Imagine that a large spacecraft equipped with a low-thrust propulsion system (possibly an ion drive) flies in formation with an Earth-threatening NEO, at as close a distance as is safely possible. If the thruster is turned off, the small gravitational attraction of the NEO will pull the spacecraft toward it until the two objects ultimately collide. But if the thruster is used to maintain a constant separation between the NEO and the spacecraft, the spacecraft's smaller gravitational pull slightly alters the NEO's solar trajectory. Given enough collision warning time and accurate knowledge regarding the NEO's trajectory, an Earth collision could again be converted to a near miss.

Human Missions to NEOs

Before we can perfect our NEO-defection techniques, we must learn a great deal about these objects. By 2020, humanity's interplanetary capabilities should be sufficiently mature to conduct NEO exploration missions.

At least for the nearest NEOs, round-trip travel times (including 1- to 2-week stopovers at the NEO) should approximate 90 days. Although NEO exploration will be more risky than roundtrip voyages to the moon, weightlessness and cosmic radiation should be less problematical than in 2- to 3-year roundtrip missions to Mars. Within a few decades, humans will almost certainly have begun the exploration of these small celestial objects.

Further Reading

To check on the recent state of Eta Carinae, consult Francis Reddy, "The Supernova Next Door," *Astronomy*, 2007;35(6):33–37. The possible connection between gamma-ray bursts and black holes is discussed by Steve Nadis, "The Secret Lives of Black Holes," *Astronomy*, 2007;35(11):29–33.

Various astronomical data sets present the orbits and properties of asteroids and comets. One of these is Arthur N. Cox, ed., *Allen's Astrophysical Quantities*, 4th ed. (New York: Springer-Verlag, 2000).

To survey craters produced on Earth by past asteroid and comet impacts, see Francis Reddy, "Illustrated: Earth Impacts at a Glance," *Astronomy*, 2008;36(1):60–61. The possible problem posed by Apophis's return visit in 2029 is reviewed in Bill Cooke, "Fatal Attraction," *Astronomy*, 2006; 34(5):46–51.

Many astronomy texts discuss the interaction of comet Shoemaker-Levy 9 with Jupiter. One is Eric Chaisson and Steve McMillan, *Astronomy Today*, 3rd ed. (Upper Saddle River, NJ: Prentice-Hall, 1999).

An excellent source to review many of the NEO-deflection concepts is T. Gehrels, ed., *Hazards Due to Comets and Asteroids* (Tucson, AZ: University of Arizona Press, 1994). The parasol concept was introduced in a paper by Gregory L. Matloff, "Applying International Space Station (ISS) and Solar-Sail Technology to the Exploration and Diversion of Small, Dark, Near Earth Objects (NEOs)," *Acta Astronautica*, 1999; 44:151-158. For current model calculations regarding the solar collector, see Gregory L. Matloff, "The Solar Collector and Near-Earth Object Deflection," *Acta Astronautica*, Volume 62, Issues 4–5, February–March 2008, pp. 334–337.

The gravity tractor is such a new concept that little has appeared as yet in refereed publications. A Web article discussing this concept is Russell Schweickart, Clark Chapman, Dan Durda, and Piet Hut, "Threat Mitigation; The Gravity Tractor," www.b612foundation.org/papers/wpGT.pdf.

NASA and other space agencies have begun planning human missions to NEOs. See, for example, Andre Bormanis, "Worlds Beyond," *The Planetary Report*, 2007; 27(6):4–10.

Mitigating Global Warming

Ken Roy, Robert Kennedy, and David Fields

All in a hot and copper sky,
The bloody Sun, at noon,
Right up above the mast did stand,
No bigger than the Moon.

Samuel Taylor Coleridge, from "The Rhyme of the Ancient Mariner"

Conservation, recycling, more efficient machines, altered lifestyles, and new sources of energy are all needed to reduce the growth rate of our greenhouse gas emissions. Unfortunately, the best that all of these actions combined can achieve is a reduced rate of emission growth. If humans are truly causing global warming by profligate use of fossil fuels, then we need to do far more than reduce the growth in our emissions; we must reduce the absolute amount of these gases in the atmosphere to levels below those of the last century. With more and more people entering the global middle class, and their commensurate use of more and more energy, it is unlikely, perhaps even impossible, that we will be able to stop global warming. We almost certainly will not be able to conserve our way back to pre-20th-century atmospheric carbon dioxide levels. We should do all that is necessary to slow down the rate at which we dump CO_2 into the atmosphere, but we are fooling ourselves if we believe we will be able to significantly reduce the amount already within it. Conservation, recycling, more efficient machines, and altered lifestyles are all Band-Aids for a problem that requires reconstructive surgery.

If we cannot eliminate that which is causing global warming, then is there anything we can do to mitigate it? The answer is yes.

There are two elements to global warming. The first, and the one about which most people concentrate their efforts, is the greenhouse gas element. Sunlight warms our planet and the greenhouse gases in the atmosphere trap some of the heat from the solar energy, keeping the overall planet warm (even at night) and making it possible for life to exist. Without a greenhouse effect, Earth would be a very cold place. Humanity's efforts are now focused on reducing the amount of heat trapped in our atmosphere by these gases.

L. Johnson et al., *Paradise Regained*, DOI 10.1007/978-0-387-79986-5_14,
© Praxis Publishing Ltd, 2010

But what about the other part of the story? What if we could reduce the amount of heat generated? What if we could reduce the overall amount of sunlight hitting Earth so that global atmospheric temperatures can return to what we consider normal? What if we were to build a large sunshade in space and reduce the amount of sunlight striking Earth, thus allowing it to cool?

Giving Earth Sunglasses

Geoengineering, or planetary engineering, is the application of technology for the purpose of influencing the properties of a planet on a global scale. Large sunshades positioned between Earth and the sun can be used to heat or cool our planet. Such an arrangement can cope with global cooling or warming here on Earth or modify the global properties of other planets to make them more Earth-like. Sunshades can also displace energy generation from an overstressed biosphere, yielding substantial economic benefits in the form of avoided costs, revenue streams, and new capabilities.

Gravity keeps our feet on the ground, satellites in orbit, and the moon in its orbit around Earth. Gravity also keeps Earth in orbit around the sun. The sun's immense mass is pulling on Earth, and if it were not for our motion in orbit around the sun, we would certainly fall into it. But Earth's mass also pulls on the sun. There is a point in space on a line connecting the Sun and the Earth where gravitational interactions allow an object placed at this location to orbit the Sun with the same angular speed as the Earth. This means that an object placed at this location would forever be between the Earth and the Sun. This point is known as the Sun–Earth Lagrange point, or L1. It is called a Lagrange point in honor of the mathematician who worked out the formulas for the physics used to describe it. In the 18th century, the Italian mathematician Giuseppe-Luigi, Count of Lagrange, finalized the mathematical theory that describes the behavior of celestial objects at the so-called Lagrange points.

If you want to see the sun–Earth L1 point, go outside during the daytime and look up at the center of the sun. It is right there, only some 1,500,000 kilometers away. (Remember to use appropriate eye protection when looking at the sun, as serious eye damage could result. Regular sunglasses are not enough.) Compare this distance with that of Earth's moon, which is about 400,000 kilometers away, or just one-fourth the distance. But, while the moon moves against the background of fixed stars over a period of a month, the sun–Earth L1 point will always be directly between Earth and the sun. Something that is placed there will obstruct some of the sunlight that would otherwise hit Earth. This action is also known as "occultation," which has

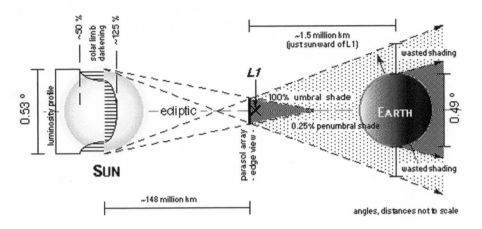

Figure 14.1. General layout of sunshades near the sun–Earth L1 region.

nothing to do with magic, though it may look like it. Note that this effect is not an eclipse either, because in general the dark part of a shadow—known as the umbra—from an object at L1 cannot reach Earth. Unless the object at the sun–Earth L1 is very large, on the order of thousands of kilometers in diameter, only the dim outer part of the shadow—the penumbra—reaches Earth. Placing enough of something at the sun–Earth L1 can significantly cut down the sunlight hitting Earth (Fig. 14.1). If we are clever, we can locate just the right amount of "stuff" there to alter the solar radiation hitting Earth to achieve any desired temperature reduction. (With a similar technique, an increase in temperature can also be achieved, because sometimes ice ages happen, too.) Also, these techniques can in principle be applied to any other planet.

Another important characteristic to note is that although the Lagrangians are called "points," they are in fact *regions* of stability. For example, the region of metastability centered around the sun–Earth L1 point above is shaped like a sausage lying about 800,000 km along the orbit (i.e., perpendicular to this page). The region's transverse dimensions (i.e., up-and-down, and left-to-right on this page) are approximately 200,000 km each. Thus the sun–Earth L1 region contains roughly 30 quadrillion cubic kilometers of space, which is of a similar order to the 200 quadrillion km^3 of cislunar space in the Earth–moon system. However, for purposes of shading sunlight on Earth, the "sweet spot" is a lot smaller, less than 1 percent of the region. Finally, not all shade is created equal; in Figure 14.1 you can see that a sunshade parked right on the sun–Earth line intercepts a higher quality of light than one parked somewhat off the axis. This phenomenon, which varies by a factor of 2.5 or so, is called solar limb darkening, and is one of the components of shading efficiency. (You can observe this phenomenon for yourself in an ordinary low-wattage incandescent light bulb. Note how the edge of the bulb looks dimmer than the center no matter what your vantage point.)

Sunshades in Space

If we are principally interested in reducing by some degree the amount of solar radiation hitting a planet and doing so with as little effort as possible, then the "stuff" we position between a planet and its star will need to have maximum surface area for minimum mass. It will need to block a lot of light and not be too heavy to launch into space. Fortunately, a technology is being developed today that might be up to the job: solar sails.

A solar sail is just what its name implies: it is a sail that propels itself using sunlight. Sunlight has no rest mass, but it does have momentum. When light is reflected from an object, the light imparts some of its momentum to that object, just like the cue ball hitting the 4 ball in a game of billiards. The momentum of sunlight is very, very small. So small, in fact, that we cannot feel it, and it is negligible on Earth where the magnitudes of all the other forces acting on us are so much larger. But in space, without air and wind and in situations where the pull of gravity from Earth or sun is also small, the tiny push from sunlight can be a significant factor in making a spacecraft move. If the spacecraft has a large, lightweight, and highly reflective sail attached, it can maneuver just about anywhere in the inner solar system without fuel, using only reflected sunlight to propel it.

But these sunshades will not be a solar sail in the traditional sense. There will be some big differences. Solar sails for geoengineering can be very heavy compared to solar sails used to transport cargo around the solar system. They will not have a payload; in a sense they *are* the payload. Also, because they will not be hauling designated freight, we do not have to make them any particular size. We can make them any size we want and we will make that choice based on ease of manufacture and perhaps other factors. For our purposes we do not really care if we have a million little sails each a kilometer square or a single large sail having an area of a million square kilometers. It is the optical properties of the sails that are critical, and will vary a lot depending on the particular engineering solution chosen. The sails will not necessarily be totally reflective mirrors, or totally absorptive blackbodies. Different parts may be mirrored, or black, or diffractive, or even transparent, or some combination. Rather than lasting just for the duration of a cargo mission, they will be built for longevity. They will be engineered to endure the harsh conditions of interplanetary space for an extended period of time. They will have to withstand high radiation fields and continual assaults by the solar wind or even the occasional solar storm, and tolerate the occasional puncture by micrometeoroids.

The sails will need to have sensors and controls and even some degree of intelligence. To maintain their assigned positions near the L1 point, they will have to vary the thrust resulting from solar radiation to counter forces that would pull them off station. Because we are talking about many thousands or possibly millions of such solar sails cruising along in orbit in close proximity to one another (much like a giant school of fish), they will also have to be social. That is, their sensors will observe their neighbors as well as the primary and the satellite, and they must be able to maneuver to avoid crashes or other conflicts, such as cutting off a neighbor's light. They must have the means to enable them to receive instructions from their engineers. We may want them to move out of position if temperatures on the planet drop too far. Not only would these sails block light like a parasol, they would use some of the light that they intercept for free station-keeping thrust, rather than conventional reaction mass squirted from an engine. Sunlight in space is effectively free, unlike rocket fuel shipped up from Earth or somewhere else, which would get expensive over the long run. The sails could easily generate enough electrical power for their onboard computers and other equipment via photovoltaics on their bright side. There is no reason why they might not generate a surplus over their own modest power requirements, and perhaps a large surplus.

To distinguish these rather specialized solar sails from the ones discussed in other chapters, we propose the term *Dyson dot*. This is a deliberate allusion to the original idea of the Dyson sphere, which was proposed by Freeman Dyson as a system of orbiting space facilities designed to completely encompass a star, thus capturing its entire energy output. The Dyson dot would be designed to block or capture only limited amounts of a star's radiation that would otherwise strike one of its planets.

A typical solar sail needs to achieve a mass-to-area ratio on the order of 10 to 20 grams per square meter to be useful. Our Dyson dots can be that light or they can be much heavier. There is no reason they could not range up into the kilogram-per-square-meter range if necessary. One factor to keep in mind is that lighter Dyson dots will experience a significant acceleration due to light pressure, and this requires that they be positioned somewhat closer to the sun, away from the traditional L1 point—perhaps 500,000 kilometers closer to the sun than the traditional L1 point. If it were heavier or less reflective, it could be closer to the usual L1 point.

To understand why this is so, we have to return to the physics of the L1 point. The planet's mass produces an attraction on an object at L1 that is exactly opposite to the gravitational attraction produced by the sun, in effect reducing the observed mass of the sun, allowing the object at L1 to have a solar-orbital period exactly equal to that of the planet. But with a solar sail we now have a third force component to deal with and that is the force resulting from sunlight hitting and to a greater or lesser extent being reflected from the solar sail or Dyson dot. This third force adds to the force of the planet's gravity on the dot, in effect moving the point of stability sunward from the usual L1 point.

Dyson Dots as a Solution to Earth's Global Warming Problem: How Much Is Enough?

We can position a school of Dyson dots at the sun–Earth L1 point, or perhaps slightly sunward of it, but how much total area do we need? How much sunlight do we need to block? This is a simple question that does not have a simple answer. Depending on future energy policies, and on various global and solar cycles, we may need to either artificially cool or warm our planet. To begin to bound the problem, it is worth noting that a 0.25 percent reduction in the sun's energy output is what is estimated to have caused the Maunder minimum. For unknown reasons, the sunspot cycle shut down between the mid-16th and 17th centuries. Astronomers refer to this period as the Maunder minimum. Historians call it the Little Ice Age. During this period, the Thames River in England froze for the first time in recorded history, crops failed, population growth stalled, and sea ice cut off Iceland from Europe. The famous Danish astronomer Tycho Brahe recorded winter temperatures 2.7° Fahrenheit below average during the last two decades of the 16th century.

If we shoot for a similar reduction, we are probably in the ballpark of what would be necessary to deal with global warming. It is important to note that while this approach could adjust the average global temperature, it would in no way address the other environmental issues associated with the continued burning of hydrocarbons.

Figure 14.1 illustrates why no stable sunshade can project a shadow spot of exactly the right size, that is, no bigger than Earth's diameter. Some shading is unavoidably wasted due to the geometry of the sun–Earth L1 system. This is the other component of shading efficiency, and the highest possible value—about 82 percent—is obtained right at L1, as depicted in the figure. However, as previously discussed, any sunshade made of a nonmagical material will have to cruise somewhat inside of the sun–Earth L1. The brighter the mirror, the further inside L1 it has to go, and paradoxically the heavier the overall mass and cost, and the less efficient the total project gets. Any real solution will be an exercise in optimizing and trading off against multiple constraints.

At the time of this writing, it appears that the gross parameters of an array of sails will vary by a factor of roughly 3: from 300,000 km^2 (about the size of the state of Arizona) and 10 to 20 million tonnes (metric tons) at the low end, to about 900,000 km^2 (somewhat greater than the entire West Coast of the United States) and 50 to 60 million tonnes at the upper end. For example, using reasonable middle values for the parasol parameters—80 percent reflectivity or albedo, mass 53 grams per square meter, positioned 2,100,000 kilometers from Earth—we would need almost 700,000 km^2 of sunshade area to achieve a reduction of 0.25 percent in the solar constant, that is, some 37 million metric tons. (This sounds like a lot, but for perspective bear in mind that the United States alone burns about one billion tonnes of coal every year. One supertanker of the many hauling petroleum around the world's oceans weighs about half a million tons fully loaded; and 37 megatonnes is roughly just 3 days' supply of crude oil.) If each Dyson dot has an area of 10 km^2, then our array, or school, would consist of 70,000 units.

As the mirror gets denser—say, 100 grams per square meter instead of 50—or the climate change problem becomes more severe, then the total mass, and hence the cost, needed to achieve the desired effect on Earth increases. If we want a 1 percent reduction in insulation, then the area, mass, and cost increase by a factor of 4; for 2 percent, they increase by 8.

On another level in the grand scheme of things, 37 million tonnes is not so much; it is the mass of a small stony-iron asteroid a mere 300 meters across—a class of rock so small that we have not seriously looked for it yet.

Infrastructure Discussions

L1 is very high up, beyond this deep gravity well we live in, and our technology is grossly unequal to the task of lofting solar sails into space. At a current launch cost of $10,000 per pound just to get into low Earth orbit (LEO), it is obvious that existing rockets are too expensive to implement this particular solution. Only a truly advanced space-faring civilization using energy on an unprecedented scale could execute such a project. However, all proposed solutions to global warming—including doing nothing, which is always an option in human affairs—are expensive or painful. For example, it has been estimated that the cost to reduce CO_2 emissions just for the United

States to 80 percent of the 1990 level could cost up to $6 trillion. The cost to bring the Chinese and Indian populations up to the U.S. standard of living while lowering their current CO_2 emissions could easily be an order of magnitude more. Perhaps becoming an advanced space-faring civilization would be the cheaper alternative.

Several proposals have been presented to implement the Dyson dots: Build a large manufacturing facility on the moon to manufacture the dots using lunar materials and then launch them from there. Build a large rail gun on Earth that could fire a spacecraft to the sun–Earth L1 point and then deploy the dots from the rocket. Move an asteroid to the sun–Earth L1 point and then establish a manufacturing facility there to make and launch the Dyson dots using material from the asteroid itself as the raw material.

Wherever it would be located, the manufacturing facility would be busy for some time. Using the previous medium-valued example, and assuming a deployment period of 10 years, we would need to launch about ten thousand tonnes of dot stuff a day. Even after we had achieved the desired reduction in sunlight, we would still have to replace units as they fail.

Energy Options

Each square kilometer at Earth orbit receives some 1400 megawatts of power. For the task of cooling Earth, it does not matter how much solar radiation the Dyson dot reflects just as long as it is prevented from getting here. If we convert the 20 percent absorbed energy to electricity at 10 percent efficiency we have 28 megawatts of power per square kilometer of mirror. In fact, solar cells—not to be confused with solar sails—have achieved efficiencies much higher than 10 percent with current technology. If we absorbed 70 percent of this energy instead of 20 percent, and converted 20 percent of that to electricity to be beamed back to Earth via maser (microwave laser), then we have almost 200 megawatts per square kilometer. Let's stick with the conservative 28-megawatt number. Keep in mind that we have got over 700,000 square kilometers of solar sail somewhat sunward from the sun–Earth L1 point. Assuming that a typical nuclear power plant produces 1000 megawatts, then our Dyson dot array can boast a power output equal to twenty thousand nuclear power plants. The United States currently has a mere one hundred or so such plants; they produce 20 percent of the country's electricity. It is clear that if we could transmit even a small fraction of the sunpower that Dyson dots intercept to Earth in the form of electricity, we would have enough to provide for all the nations of this world, with plenty left over. Certainly displacing polluting power generation off-world would be a good thing, and a net benefit to the ecosystem, since terrestrial carbon burners waste two thirds of their fuel's energy due to basic thermodynamic inefficiency.

The solar wind is composed mainly of hydrogen and helium. Some of the helium is the isotope helium-3. Our mirrors will be exposed to the solar wind for many years. They could be designed to capture these particles and then return to a central processing facility at the end of their life span for recycling. Hydrogen, helium, and especially helium-3 could be very valuable to a space-faring civilization. Helium-3 holds great promise as a fuel for fusion reactors that can be easily transported to Earth in an extremely compact hence valuable form.

Looking Backward: Aesthetics and Agriculture

Suppose at some time in the future that we have built a major manufacturing facility on the moon and it has been building and launching mirrors for 10 years. Furthermore, suppose that the scientists and engineers then tell us we have reached our goal and have finally positioned enough mirrors at the proper point to reduce global temperatures to United Nations–specified levels. It is a world holiday! You go outside and look up at the sun, using the appropriate eye safety gear provided. What do you see? Nothing special. It is not possible to detect even 700,000 square kilometers of Dyson dots floating somewhere inside the sun–Earth L1 point against the backdrop of the sun itself, because their dark umbras never get to you on the ground. Nor are you likely to feel the effect on your skin, any more than you feel the shadow of a bird flying overhead. Using special devices designed to image sunspots, you might see a strange cluster of black dots in the center of the sun if you were willing to make the effort. To most people going about their normal affairs on this planet, however, they would be invisible, like all good infrastructures. You shrug and decide to go seek more interesting sights.

The farmers in this future have pointed out at the start of the Dyson Dot Project that a quarter-percent reduction in sunlight means a quarter-percent loss in crop yield. But the farmers, no less obstreperous than the ones today, have been assured that no such shortfall would happen. It turns out that plants are not very efficient at using the sunlight provided to them. Chlorophyll uses only narrow bands of light in the blue-purple and orange-red regions of the spectrum. They paradoxically do not use green at all, which is right near the *peak* (most abundant energy) of the solar spectrum.

Green light is reflected away, which is why green plants look that way. So a number of Dyson dots have been modified to augment ground-side agriculture by converting some of the energy they collect to particular frequencies used by chlorophyll. This supplemental light is beamed to Earth using weak lasers carried on special dots. There is even talk of augmenting these frequencies above natural levels, but only for Earth's agricultural regions. Best of all is the continuing dividend that the Dyson Dot Consortium has been paying to the citizens of Earth via power sales and reduced insurance premiums. (The economic justification is a topic for another book, however.)

A Bigger Perspective

Earlier we mentioned other worlds and other places. For example, if we wanted to reduce the solar radiation hitting our neighbor Venus to levels that are normal on Earth, as opposed to the venuforming of Terra, which seems to be going on right now, we would need to block about half (48 percent exactly) of the incoming sunlight. This can also be done using Dyson dots located at the sun–Venus L1 point, but the level of effort would be approximately 200 times as great as that required to address the global warming problem right here on Earth. We recognize that the terraforming of Venus would involve much more than just adjusting its solar constant, but that task would have to be part of the overall terraforming effort. If we were terraforming Mars, on the other hand, we would do just the inverse—double the Martian insulation with off-axis dots. Projects like this could take thousands of years without quasi-magic methods like self-replicating nanotechnology. Nevertheless, how much would a second Earth in this Solar System be worth to the human race? Survival trumps the ordinary calculus of economics.

Given the Anthropic principle, none of these considerations are unique to our home system. Physical laws are the same for everybody; thus it is reasonable to suppose that any intelligent beings may utilize techniques we would recognize. Dyson dots as described in this chapter can be used to increase or decrease the solar constant and to some extent modify the color of light hitting the target planet. They can be used as power stations and as resource collection vehicles. If we were interested in terraforming a planet, this would be a handy technique to employ. If global warming is a real threat, can we deploy this tool in time to make a difference? Maybe, but certainly not without cheap and reliable access to space, and certainly not without the national or international will to do great things. It is clear that building and using Dyson dots would create a set of mutually reinforcing capabilities that would each be valuable, even indispensable, to a space-faring civilization. Solving the climate change problem here on Earth may be just the thing to bootstrap the human race to a new level of existence.

But keep in mind that although we may find it too daunting at present, beings in other solar systems may already have employed this technology. Perhaps researchers at the Search for Extraterrestrial Intelligence (SETI) should begin to look for the occasional flash from distant Dyson dots. They are designed to reflect a lot of energy, and it must occasionally be seen by outside observers. These other beings may not want to talk to us, but it would somehow be comforting to know that they are making themselves comfortable on a distant world.

Further Reading

For further information on the topics discussed in this chapter, we recommend the following: M.J. Fogg, *Terraforming: Engineering Planetary Environments* (New York: Society of Automotive Engineers, 1995); Louis Friedman, *Starsailing: Solar Sails and Interstellar Travel* (New York: John Wiley, 1988); Peter Glaser, *Solar Power Satellites* (New York: Arthur D. Little, 1968); Thomas Hayden, "Curtain Call," *Astronomy* magazine, edited by B.B. Gordon (January 2000):45–49; Christopher C. Kraft, *The Solar Power Satellite Concept* (1979); Colin R. McInnes, *Solar Sailing: Technology, Dynamics and Mission Applications* (Chichester, UK: Praxis, 1999); Alan S. Manne and Richard G. Richels, "CO^2 Emission Limits: An Economic Cost Analysis for the USA," *The Energy Journal*, edited by Leonard Waverman, Washington, DC, 11 (April 1990): 51–74; E. Mallove and G.L. Matloff, *The Starflight Handbook* (New York: John Wiley, 1989); Gregory Matloff, *Deep-Space Probes*, 2nd ed. (New York: Springer-Praxis, 2005); G.L. Matloff, L. Johnson, and C Bangs, *Living Off the Land in Space* (New York: Springer-Praxis, 2007); K.I. Roy, "Solar Sails: An Answer to Global Warming?" in *CP552, Space Technology and Applications International Forum – 2001*, M.S. El-Genk, ed. (2001); Ken Roy and Robert G. Kennedy, "Mirrors and Smoke: Ameliorating Climate Change with Giant Solar Sails," *Whole Earth Review*, edited by Bruce Sterling (Summer 2001): 70; John P.W. Stark, "Celestial Mechanics," in *Spacecraft Systems Engineering*, Peter Fortescue and John Stark, eds. (Chichester, UK: John Wiley, 1991), pp. 59–81; G. Vulpetti, L. Johnson, and G.L. Matloff, *Solar Sails: A Novel Approach to Interplanetary Travel* (New York: Springer-Praxis, 2008).

Settling the Solar System   15

> Water, water everywhere,
> And all the boards did shrink;
> Water, water everywhere
> Nor any drop to drink.

Samuel Taylor Coleridge, from "The Rhyme of the Ancient Mariner"

The thought is not new, but it is profound and it can help guide humanity as we deal with the myriad environmental challenges facing us at the beginning of the 21st century: we all inhabit one giant spaceship on a voyage together through the dark emptiness of space. We call it "Spaceship Earth."

Clearly visible from the hotels in Orlando, Florida, is Disney's Spaceship Earth at the Epcot Center (Fig. 15.1). When visiting the dome, tourists experience the history of human communication, from the dawn of humankind to today. For many, seeing the dome reminds them of the interconnection we have with each other and with our home planet—a more far-reaching outcome than is intended.

As we learn to be better stewards of Earth and thereby protect it for habitation and use by future generations, we will also be learning how to better extend the human presence beyond Earth and into the nearby solar system. The reverse process is also true. Many of the technologies being developed to support human life beyond Earth may have application here at home as well. This chapter examines the essential parts of a self-sustaining outer space habitat, and discusses how this relates to similar processes on Spaceship Earth.

The essential elements of a crewed space mission include Earth-to-orbit transportation, transportation in space and away from Earth, energy, food, water, waste disposal or recycling, and the commodities of everyday life. We will not dwell on the transportation systems required for space exploration. For more complete background on both Earth-to-orbit and in-space transportation systems, see the authors' previous books, *Living Off the Land in Space* (New York: Springer-Praxis, 2007) and *Solar Sailing: A Novel Approach to Interplanetary Travel* (New York: Springer-Praxis, 2008). We

L. Johnson et al., *Paradise Regained*, DOI 10.1007/978-0-387-79986-5_15,
© Praxis Publishing Ltd, 2010

Figure 15.1. Spaceship Earth at the Disney World Epcot theme park reminds visitors of the giant spacecraft we all inhabit. (Courtesy of Katie Rommel-Esham.)

will discuss the other issues cited above, as all of them are linked to terrestrial analogs with which they have much common ground.

Water

As we explore space, it is vital that we find water. We humans need it not only for drinking, but also for bathing, cooking, and for use in our rockets. The Apollo astronauts took with them enough water for the round trip to and from the moon, with some in reserve. The International Space Station recycles some of its water, but must still be regularly resupplied from Earth. Any long-term settlement will need a supply of water and a way to recycle it.

Water Facts
- Approximately 70.8 percent of Earth's surface is covered by water— roughly 1,260,000,000,000,000,000,000 liters.
- 98 percent of the water on the planet is in the oceans.
- 1.6 percent is in the polar ice caps and glaciers.
- 0.36 percent is found underground (in aquifers and wells).
- 0.036 percent is found in lakes and rivers.

- By weight, our bodies are approximately 62 percent water.
- You cannot live more than a few days without water, depending on the temperature, your activity level, and your general state of health.
- Water can be separated into its constituents, hydrogen and oxygen, and used as rocket fuel.

Fortunately, nature provides us with water throughout the solar system. The early Earth is thought to have received its water from frequent comet impacts. Other planets and moons experienced similar impacts and therefore should have water as well. We now know that there is water on Mars, locked mostly in subsurface ice and in the planet's polar ice caps. The moon may have water in the perpetually shaded craters at its south pole. Based on recent observations, the main belt asteroid Ceres may be 25 percent water. The many comets that periodically visit the inner solar system are composed mostly of water ice.

Just because there is water at potential settlement destinations does not mean that we can use it profligately. As discussed in other chapters, water is a valuable resource and one that must be used wisely with an eye toward recycling. There simply is not enough of it out there, nor will what is there be easily accessed. Nothing accomplished in space is easy and every attempt must be made to use and reuse the resources already "in hand."

NASA has been working on the water problem for years and is implementing a water recycling system on the International Space Station (ISS). Otherwise, NASA and its partners would have to launch about 18,000 kilograms of water annually just to keep the station functioning. The water recycling system, technically known as the Environmental Control and Life Support Systems Water Recycling System (ECLSS-WRS), reclaims and prepares for reuse water from the crew's urine, showers, and that which is exhaled into the air. (When on the ISS, don't ask where that cool drink of water came from!)

Water on Earth is recycled, although we do not usually think of it in that way. On Earth, water passes through our bodies and is put back into the environment. Microbes in the soil break down our wastewater and convert the waste part of it into nutrients that plants can use and absorb to grow. These microbes and the soil act as filters, purifying the water for later consumption by someone or something else. Some of our wastewater evaporates and is filtered by atmospheric processes before it falls back to the ground as rain or snow.

Scientists studied nature to design the systems our space explorers will use to recycle water. The techniques developed for space are now being used to help improve terrestrial water treatment plants' efficiency and efficacy.

Air

Astronauts need air to breathe when they are in space. The Earth's envelope of breathable air extends only a few kilometers upward and then thins to almost nothing at 100 kilometers.

When walking, the average man breathes approximately 30 liters of air per minute. The average woman requires 20 liters per minute. For both sexes, the amount required increases dramatically when exercising.

Air Facts

- The air we breathe is approximately 78 percent nitrogen and 21 percent oxygen. Carbon dioxide (CO_2) is present in trace amounts.
- 75 percent of Earth's atmospheric mass is less than 11 kilometers above its surface.
- Trees make their own food from the CO_2 in the air and release oxygen for us to breathe.
- One acre of forest can produce enough oxygen to sustain 18 people each day.
- You cannot live more than just a few minutes without air.

Where will future space explorers get their air? The air we breathe here on Earth is recycled. The oxygen in the air is the part that we consume and turn into CO_2 during respiration. We breathe in air, mostly composed of nitrogen and oxygen, and we breathe out air composed of nitrogen and CO_2. By removing the carbon dioxide and replenishing the oxygen, we can maintain a breathable atmosphere. The problem for space explorers, therefore, is twofold: getting rid of the CO_2 and replacing it with oxygen.

Onboard the ISS and in most long-duration human space flights, CO_2 is removed from the air and dumped overboard. The oxygen is replenished with resupply from Earth. This clearly will not work for long-duration missions in deep space. Where, then, will explorers get their oxygen? The answer to this question depends on where the explorers are going.

As mentioned above, the moon may contain water in the shadowed craters near its solar pole. Without sunlight, the remains of long-ago comet strikes may lie dormant deep within these craters, providing a large supply of ice for future explorers. Water and ice are composed of hydrogen and oxygen. These chemicals may be separated when an electrical current flows through the water in a process called electrolysis. Electrolysis is not new. Since its discovery in 1800, electrolysis has been used in many industrial processes and may be used in the future to mass-produce hydrogen for Earth-based fuel cells.

Anywhere our explorers and settlers encounter water, they can use the energy of sunlight to produce electricity and, through electrolysis, oxygen. But what about places were there is little or no water? When in deep space, will explorers be able to find other sources of oxygen?

Again, our nearest neighbor, the moon, has additional resources that might be used to produce oxygen. Many lunar rocks contain the mineral ilmenite, which is made of iron titanium oxide. Thanks to the Apollo missions, we know that ilmenite is plentiful on the moon. To release the oxygen trapped within the rocks, it is in principle as easy as adding hydrogen and heat (from sunlight). The technologies for accomplishing this are in hand and could fairly easily be adapted for use in space. For example, an inflatable solar concentrator (think of it as a balloon shaped like a magnifying glass) could concentrate sunlight on a sample in a hydrogen-filled pressure vessel, heating the rock to greater than 1000 degrees. The process would produce iron, titanium oxide, and water. The water could then be converted to oxygen and hydrogen.

Earth is very efficient at recycling CO_2, though perhaps not efficient enough. The rate at which human industrial civilization is pumping CO_2 into the atmosphere appears to exceed the rate at which the currently established natural processes can remove it. Application of proposed carbon sequestration and reuse technologies could benefit both Earth and our deep space explorers. For example, captured CO_2 from a power plant or from astronaut respiration could be used to produce methanol. Methanol, an alcohol, could be used as a fuel for transportation or electricity-producing fuel cells. Today, methanol is synthesized directly from fossil fuels, but tomorrow it could be made from the waste produced by burning fossil fuels or from the very air we exhale.

Power

The key to space settlement is power. Power is required for the transportation systems that will get our settlers into space and to their final destinations. Power will be required to keep them from getting too warm or too cold. It will be needed to run their machines, produce and recycle water, replenish the oxygen in the air they will breathe, and to power their scientific instruments. It will be needed in every facet of space exploration and settlement and cannot be taken for granted as early Earth-bound explorers could do, knowing as they did that the sun would rise the next day to provide them with light and heat. In space, taking advantage of the power from space will not be as easy as merely basking in the sunlight from one day to the next.

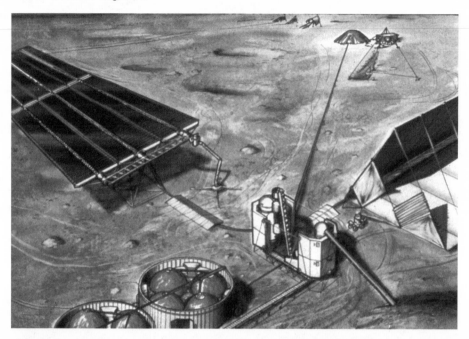

Figure 15.2. Solar power stations on the moon may provide much of the power required for future human settlements. (Courtesy of the Space Studies Institute.)

Many of the technologies our space settlers will use to generate power are closely analogous with their terrestrial counterparts. For settlements near the sun, which is all we can expect for at least the next 100 years or so, solar power should be the primary option. On a small scale, our explorers can extend arrays of solar cells outward from their habitats to gather as much sunlight as possible for direct conversion to electricity. Near Earth, the energy density of sunlight is 1368 watts per square meter. With no cloudy days, space settlers should be able to convert this perpetual source of energy into electricity with close to 40 percent efficiency.

Solar cells are primarily made from silicon, which is abundant on both Earth and the moon. On the moon, solar cells might therefore someday be made from the lunar dirt in much the same way they are made from sand here on Earth. Large solar arrays can be deployed on the surface of the moon to generate most of the power needed for a human settlement. An artist's conception of a solar array "farm" near a lunar settlement is shown in Figure 15.2.

Unfortunately, the length of time the sunlight is available depends on the location, and some locations are sunnier than others. In low earth orbit, where the first space industrial parks may some day be constructed, orbiting facilities will experience darkness 40 to 50 percent of the time as they enter and travel through Earth's shadow. (Solar cells do not produce any power when there is no sunlight.) To keep the power on during these frequent eclipses, technologists are developing better batteries that can store and release large amounts of power repeatedly with high efficiency. One of the most promising of these new batteries is called a flywheel, which is a mechanical device that stores energy in the rotation, or spin, of an internal wheel. When power is put into a flywheel, the rate of spin increases. When power is removed, the spin rate decreases. Unlike virtually all conventional batteries, including fuel cells, no chemical processes are involved. Flywheels are strictly mechanical, and they can theoretically store very large amounts of energy with frequent charge and discharge cycles, suffering little or no loss of efficiency as they age.

The power system of an Earth-orbiting industrial park would have to generate roughly twice as much power as it needs during the sunlit portion of its orbit so that it can store 50 percent of the power generated by spinning up its flywheel batteries. When the park enters Earth's shadow and the power from its solar cells drops to zero, the energy stored in the flywheels is tapped to keep the station running until sunlight returns. This spin-up and spin-down cycle is repeated every time the station completes an orbit.

Flywheels could also be used on the moon, where the day/night problem is much more severe. The moon's rotation is much slower than Earth's. As a result, the length of a lunar day is much longer than its terrestrial counterpart. Instead of 24 hours, the lunar day is 28 Earth days long, with almost any given lunar location experiencing approximately 50 percent of that time in darkness. When you depend on the sun for power, a 14-day "night" is a long night indeed. The flywheel energy storage system required for the moon will be very large.

The situation faced by space habitats needing supplemental power when they are not able to generate it on their own is analogous to that faced by terrestrial power stations during their peak loading periods. Consider, for example, the power demand placed on utilities in the American south during the summer months. The peak demand for power occurs in the mid-afternoon when the sun is high and office buildings, stores, and factories are using their air conditioners. This is also the time that some workers are starting to go home and are turning on their home air conditioners. The demand for electrical power during this time often outstrips the ability of the utility to generate it. This results in the importing of expensive power from outside the local grid, thereby raising costs. Alternatively, if no supplemental power is available, voluntary or involuntary brown-outs might occur.

If large flywheels could be installed to store power generated during times of lower power demand (for instance, at night), and release it during these daytime peaks, then the need for expensive non-utility-generated power could be reduced and the occasional brown-out avoided. On-demand power and efficient power storage are required for both space and Earth.

Finally, the space solar power stations described in Chapter 11 could beam energy to Earth orbiting or lunar habitats as required. Once a space-based power infrastructure is in place, the uses for it will multiply.

Radiation Protection

Space is a harsh and deadly radiation environment from which humans need protection—here on Earth and in our space settlements. The sun produces far more radiation than we can see using our eyes. The fusion-driven furnace within the sun is producing x-rays, gamma rays, and highly energetic protons, electrons, and alpha particles—all of which are permeating the near-space environment in such quantities as to pose a threat to human health from near the sun to Earth, Mars, and beyond. Without some form of shielding (from this radiation), life cannot survive.

Earth does an excellent job of protecting us from most of this deadly radiation. The same magnetic field that makes our compass needles point toward the North Pole acts as a *Star Trek*–like deflector screen, diverting and trapping much of the deadly radiation pummeling our planet on a daily basis. We can even observe some of this trapped radiation as it enters our atmosphere and interacts with it near the North and South Poles; we call it the aurora borealis. The aurora is nothing more than high-energy charged particles traveling along Earth's magnetic field careening into the dense atmosphere, ionizing it at high altitudes and causing atmospheric oxygen to glow.

While Earth's magnetic field will not stop all of the electromagnetic radiation from the sun, our thick atmosphere absorbs or attenuates much of it. Only visible light and some of the ultraviolet can get through the kilometers of air that separate us from the vacuum of space. Fortunately, the chemical composition of the atmosphere allows it to absorb much of the ultraviolet light through its interaction with ozone. Ozone is a naturally occurring form of oxygen produced in the upper atmosphere. In the 1970s, satellite data showed that Earth's ozone layer was depleting, producing potentially dangerous ozone "holes" over some parts of the globe. These holes allowed more ultraviolet light to reach Earth's surface, potentially causing increased incidences of cancer. Many scientists believed that certain chemicals commonly produced and used in industrial and commercial processes caused the depletion of the ozone layer. The body of evidence supporting this view was enough to convince politicians that the use of these ozone-depleting chemicals should be reduced. Laws to reduce their use were enacted, and the slow recovery of the ozone layer is thought to be occurring today.

Space explorers will not have the benefit of kilometers of atmosphere, magnetic field deflector screens, and ozone layers to protect them. Instead, they will have to rely on artificial shielding and advanced warning of impending solar storms, which are increased periods of high radiation, so that they can take shelter and not experience potentially lethal exposure to the solar radiation. Interestingly enough, the same solar storm warning system that space settlers will require is in use today, providing spacecraft operators and utility companies with advanced warning of solar storms so that they do not experience catastrophic failures themselves.

Before one can understand how a solar storm can cause the lights to go out on Earth, some aspects of physics that are out of the realm of most people's everyday experiences must be explained: The burst of radiation that is a solar storm is composed, in large part, of charged particles: protons, alpha particles, and electrons. These moving charged particles generate a strong and massive magnetic field. When the storm encounters Earth, it tends to push Earth's magnetic field inward toward the surface, lowering the altitude of the magnetic field lines dramatically.

When charged particles move through a magnetic field, they experience forces acting upon them. The reverse is also true. A moving magnetic field induces a current flow in a wire. A current is nothing more than a flow of electrical charges through a wire or some other conducting medium. Electrical utility wires (particularly those hanging from telephone poles at northern latitudes) feel the effect of the solar storm as Earth's magnetic field is compressed toward Earth, changing in intensity with time. This changing magnetic field induces current flow in the wires and voltage differences at the various grounding points within the power grid, creating spurious currents that knock out transformers and otherwise disrupt or shut down the transmission of electrical power.

This is a real effect and it has happened. On March 13, 1989, a solar storm sent the Hydro-Quebec, Canada, power grid, which serves more than 6 million people, into a blackout. These storms are such a threat that utilities monitor space weather conditions so that they can have repair crews available to fix the inevitable damage to their infrastructure when such storms are predicted to arrive.

To provide at least some warning of impending solar storms, the National Oceanic and Atmospheric Administration (NOAA) and NASA placed the Advanced Composition Explorer (ACE) spacecraft at one of the Earth–sun

Lagrange points. A spacecraft placed here will likely remain unless some outside force acts upon it. The regions are not 100 percent gravity or disturbance free, so some spacecraft propulsion is required to remain within them. The fuel required, however, is much less than would be needed should these regions not exist.

The ACE spacecraft detects a solar storm when radiation from an associated event on the sun strikes its detectors; the spacecraft then sends a radio signal back to Earth, also traveling at the speed of light, telling satellite operators and even earth-bound electrical power utilities that a storm is coming and to get ready. The light from the sun, and subsequently the radio transmission signaling the impending storm, reach Earth about one hour before the ionizing radiation because light travels faster in the vacuum of space than do the charged particles originating from the sun. This same sort of spacecraft will send a warning to our space settlers, urging them to take shelter.

Space settlers must also contend with another form of radiation: galactic cosmic rays. These electrically charged particles are accelerated to near-light-speed velocities by cosmic electromagnetic fields. Because only about two dozen humans have thus far voyaged beyond the protective confines of Earth's magnetic field, we have little data regarding the safe threshold for human exposure to these particles. The best methods of protecting against galactic cosmic rays are either to equip the space settlement with a massive, thick shield of rock or soil gathered from celestial bodies, or to generate an artificial magnetic field. As humans travel once again to the moon (and other solar-system bodies), space mission planners will learn a great deal more about this radiation source.

As we learn how to explore space, we will learn more about how to improve life on Earth. From taking the technologies developed to purify water on the International Space Station and adapting them for use in terrestrial water purification systems to the new technologies for capturing and sequestering carbon dioxide, the tools we will need for space development are increasingly becoming the same tools we need here on Earth.

Paradise Regained: An Optimistic Future

16

Oh yesterday the cutting edge drank thirstily and deep,
The upland outlaws ringed us and herded us like sheep,
They drove us from the stricken field and bayed us into keep;
   But tomorrow,
   By the living God, we'll try the game again!

John Masefield, from "Tomorrow"

"Gloom and doom" is easy to sell. Simply watching the evening news fills one with a sense of what is bad in the world: murder, child abuse, an increase in the price of gasoline, water rationing due to drought, tension among nation states arising over a religious or economic disagreement, and so on. Listening to the prognostication of the coming age of scarcity and environmental disaster makes for great drama and feeds what seems like an increasing sense of global pessimism.

Yet, while the gloom and doom sells newspapers and movies, progress marches onward: more people today are free than at any time in history, the standard of living for the average person on Earth has steadily increased, and many diseases that once ravaged the human population no longer exist. The human condition is undoubtedly improving, and we should revel in it.

However, the warning signs for potential danger are real. Leaving aside the hotly debated question of the degree to which we humans are contributing to climate change and global warming, we posit that all humans should be concerned about the Earth's environment and its overall health. It is our one and only home for the foreseeable future and we should do all that we can to protect it and preserve it for future generations. We should personally reduce our trash, recycle what we can, and use sparingly what we cannot. We should recycle, reuse, and efficiently consume only as much as we need, and we argue that this can be done without causing lowered living standards if we make our investments wisely and choose a path that leads to increased prosperity and living conditions for all humanity. It is here that space advocates can work with environmentalists to make that prosperous and Earth-friendly future a reality.

L. Johnson et al., *Paradise Regained*, DOI 10.1007/978-0-387-79986-5_16,
© Praxis Publishing Ltd, 2010

By working to decrease the cost of launching payloads into space, we will enable that constellation of space solar power satellites to be built around Earth, beaming downward nearly infinite energy to provide power for a prosperous humanity. With increased energy availability has historically come an increase in the standard of living. Power provided from space will allow the eventual closure of oil- and coal-burning power plants, reducing the level of atmospheric pollution and global CO_2 emissions. At night, people around the world will have a visible sign of progress gleaming in a ring around the world as they look skyward. Reflecting sunlight from their massive solar arrays, the solar electric power stations will serve as an inspiration for the next generation as they envision where we will next go in our exploration and utilization of space.

Residents near the world's space launch sites will see the flow of material rocket skyward toward a destination between Earth and the sun as the world's sunscreen is erected there. Stopping a mere 0.25 percent of the light reaching Earth will reduce the overall energy input to the planet, providing an offset to the excess heat trapped near the surface from profligate greenhouse gas emissions in the 20th and early 21st centuries. Earthbound residents will have reduced their net greenhouse gas emissions to near zero from recycling, reuse, and space solar power. But to actually reduce the levels of atmospheric CO_2 to preindustrial levels will be well beyond our reach for quite some time. To buy us the needed time, the sun shield will not only halt increases in global temperatures, but also reduce them to levels experienced prior to the industrial revolution.

Privately funded space ventures will be visiting nearby asteroids to mine their resources and bring them back home for use on Earth. Instead of strip mining entire mountains, re-greening of previously mined mountains can begin. The need for raw materials will not decrease as prosperity increases; rather, it, too, will increase. A steady supply of raw materials will be shipped toward space or lunar-based processing and manufacturing facilities to meet the demand of Earth-bound consumers. These consumers will use the products made from off-world resources and then, with ever-increasing efficiency and simplicity, recycle them for reuse.

Yet other missions will be launched to potentially Earth-threatening asteroids to deflect them from collision with the Earth to more benign orbits for either mining or to be never seen again. The specter of global catastrophe from asteroid collision will be eradicated from the list of threats to the human race.

Robots in Earth-orbiting factories or those on the moon will accomplish more and more of our manufacturing. Industrial sites that once marred the landscape can be taken apart and the land reclaimed. At first only the most environmentally hazardous facilities will be located off-planet, with their often-unavoidable toxic by-products forever banished to the vastness of empty space or on a collision course with the sun rather than with a living, breathing Earth. As accessing, living in, and working in space become more commonplace, more industry will be moved there in order to be closer to their primary source of raw materials (asteroids) and power (the sun). Eventually, more and more people on Earth will not have to live and love in the shadow of smokestacks and toxic dumps.

All the while, a small fleet of spacecraft will be circling Earth, monitoring the environment, and providing a continuing assessment of our progress toward reclaiming Earth and returning it to be what it should be for humanity: a place to live, a Paradise Regained.

Afterword

Why Space Advocates and Environmentalists Should Work Together

Aerospace is not usually considered verdant. Jet aircraft emit tons of carbon dioxide into the atmosphere, the companies that make them are usually defense contractors, and nobody considers weapons of war, necessary or not, to be green. These companies are in business to make money for their shareholders and will only be green if it somehow benefits their bottom line or if law or mandate requires them to be so. This reality, and there are exceptions, also taints those involved in space exploration and development. But painting the entire canvas with the same brush is not only grossly unfair, it is also incorrect.

Many people who devote their lives and careers to the exploration of space are not in it for the money. Sure, they need to earn a living like everyone else. But that is not why the authors of this book chose their careers. Greg Matloff is an astronomer and astronomy professor. He lives, breathes, studies, and teaches about the heavens. He does this, in part, because studying the universe tells us something about ourselves. As we have explored the universe around us, we have learned how precious our planet Earth truly is.

C Bangs explores archetypes of Earth and depicts cosmological elements from an ecological and feminist perspective. One premise of her work is that we are part of Earth and all the elements of our bodies at one time were within a star. We contain both systems within us. Her art in a wide variety of media is informed by mythology and the hope for human evolution. Space is not only about individual heroics or national pride. Ultimately, it deals with the expansion of consciousness and the survival of the human species and other terrestrial life forms.

For example, by studying the climate of Mars, we have something to which we can compare Earth's climate. This provides us with another data point, and that is vitally important because it is difficult, if not impossible, to explain anything generally with only one example. An analogy would be trying to draw a line using only one point. You are free to draw such a line in any direction within a 360-degree angle. Making conclusions about any specific line that one draws as being "correct" is therefore impossible. With two points, you may draw a line. It may still not accurately describe the system, but it will likely be closer to correct than any other of the infinite set of lines that could have otherwise been drawn through a single point.

Les Johnson is a NASA physicist and manager. He dreamed of working for NASA since he was a 7-year-old boy in Ashland, Kentucky, and watched on television as Neil Armstrong set foot on the moon. He knew then that he wanted to be part of that great adventure and that he had to become a scientist in order to do so.

The three of us are not unique nor are we in the minority among those who work in the field of space exploration and science. Most of our colleagues with whom we have discussed career motivations ("Why did you study science and become an astronomer/physicist/engineer?") share our fundamental love of knowledge and passion to explore, develop, and use space for the betterment of humankind. And this passion is not limited to scientists and engineers.

Businessman Robert Bigelow is using the fortune he earned in business to foster space development. He is the founder and owner of Bigelow Aerospace near Las Vegas, Nevada, which is building what may be the world's first orbital hotel. Bigelow has already flown subscale prototypes in Earth orbit and has plans to loft full-scale pressurized modules within the near future.

Another businessman, Elon Musk, is using his dot.com wealth to develop a new generation of inexpensive rockets to carry machines and eventually people to space at a lower cost than is currently possible.

Environmentalists share the vision of protecting and preserving life on Earth. By working together we stand a much better chance of being able to make this vision a reality.

As we gaze skyward on a starry night, we are struck by how immense and seemingly lifeless the universe appears to be. If we are not alone among the stars, then we are among a very few civilizations likely separated by an abyss that will be impossible to cross. Hence the imperative that we preserve that which makes us unique: we are part of a living planet. Exploring and developing space will help us make this a better world.

Les Johnson, Madison, Alabama
Gregory L. Matloff, New York, New York
C. Bangs, New York, New York

Index

Advanced Composition Explorer
  Spacecraft, 165
Agricultural Revolution, 30–2
Agricultural Revolution, 30–2
Air recycling, 160–1
amphibians, 23
Anthropic Principle, 154
Apollo 8 Lunar Mission, 11920
Arcopolis, 446
Arctic Sea Ice, 53, 55, 126
Ares-I Rocket, 96, 97
Ares-V Rocket, 96, 97
Asteroid Apophis, 137
Asteroid Eros, 94
Asteroid Ida, 136
Asteroid Itokawa, 95, 137
Asteroid mining, 98–9, 170
Asteroids, 12, 14, 61
Attenborough, David, 61

Bayeux Tapestry, 99–100
Bigelow, Robert, 176
Biodiversity, 59–65, 123
Biofuel, 63, 76, 107–8
Bronze Age, 40, 81–2
Brown's Ferry Nuclear Plant, 69
Buffalo Mountain Wind Farm, 74

Carbon dioxide, atmospheric, 53, 161
Chad, Lake, 128, 129
Climate, 50
Climate change, 30, 49–56
Coleridge, Samuel Taylor, 145, 157
Comet Neat, 101

Comet Shoemaker-Levy 9, 138
Comet Swift-Tuttle, 5
Comet Wild, 2, 17
Comets, 99–101, 135
Corn, Bt, 123
Cosanti Foundation, 46
Cretaceous Era, 25
Cretaceous Extinction Event, 25

DDT, 62
Deinococcus radiodurans, 7
Desertification, 128–9
Devonian Era, 23
Dunne, John, 107
Dyson Dot, 148–152
Dyson Sphere, 149
Dyson, Freeman, 149

Earth rise (photograph), 120
Electrolysis, 160
Electromagnetic radiation, 2–4, 120
Energy consumption, annual, 68, 108
Environmental Protection Agency
  (USA), 123
Evolution, 21–26
Extinctions, mass, 24, 60–1, 134

Frost, Robert, 49

Gaia Hypothesis, 65
Genetic engineering, 123
Geostationary Orbit, 110–11
Geothermal energy, 75–6
Glaser, Peter, 109, 155

L. Johnson et al., *Paradise Regained*, DOI 10.1007/978-0-387-79986-5,
© Praxis Publishing Ltd, 2010

Gravity Tractor, 141

Heat engine, 34–35
Helium-3, 91–93, 153
Huag, Gerald, 50
Huntsville, Alabama, 4
Hydropower, 72 , 73
Hyperspectral imaging, 124

Ice Age, 30
Iguazu National Park (Argentina), 122
Intergovernmental Panel on Climate
    Change, 53
International Space Station, 111, 112,
    113, 114, 122, 158
Iron Age, 32–3, 40, 82–3

Jupiter, 14, 102

Kuiper Belt, 16
Kuiper Belt Objects, 16, 135–6

Lagrange Points, 146–7
Lake Guntersville Dam, 72
Lanier, Lake, 128
Lewis, John, 103
Little Ice Age, 51–52
Living Off the Land in Space, 98, 157
Lovelock, James, 65

Magnetohydrodynamics, 34
Magnetosphere (Earth's), 4–5, 164
Malthus, Thomas, 43
Margolis, Lynn, 65
Mars, 13–14, 102
Masefield, John, 169
McDevitt, Jack vii–ix
Mercury, 13, 90
Microwave Power Beaming, 109, 113
Middle Ages, 40, 83–4
Monarch Butterfly, 63
Moon, 6, 7, 91, 161
Musk, Elon, 176

National Aeronautics and Space

Administration (NASA), 96–7, 101,
    112, 114, 123, 159
National Space Society, 110, 116
Near Earth objects, 94–9, 136–42
Neolithic Era, 40, 62, 81
Neptune, 1516, 102
Neutrinos, 12
Nuclear power, fission, 69–71
Nuclear power, fusion, 11, 71–2, 91–3

O'Neill, Gerard K., 44,
O'Shaughnessy, Arthur, 79
Old Faithful Geyser, 75
One Planet Many People: Atlas of Our
    Changing Environment, 121, 122, 129
Oort Cloud, 16, 138
Overview effect, 119, 130
Ozone, atmospheric, 164–5

Paleolithic Era, 40, 79–80
Perseid meteor shower, 5
Pluto, 16
Population, human, 39–47

Remote sensing, 120
Renaissance, The, 84–5
Robinson, Edwin Arlington, 67
Roskill Information Services, 95
Rossetti, Dante Gabriel, 11

Sandburg, Carl, 119
Saturn, 15, 102
Sea ice, 53
Sea level rise, 124–6
Shakespeare, William, 133
Shelley, Percy, 21
Small Expendable Deployer System, 114
Solar concentrator, 97
Solar constant, 51
Solar power output, 107
Solar sail, 139–40, 148–51
Solar storm, 165–6
Solar system formation, 11–18
Solar wind, 4, 90
Space solar power, 91, 107–16, 162–3

Spaceship Earth, 157–8
Sun, 107, 108,
Supernova, 134

Temperature, of the atmosphere, 126–7
Thermodynamics, First Law of, 34, 67
Thermodynamics, Second Law of, 34–6, 67
Titan, 21, 22
Transgenic crops, 123
Tunguska event, 137
Turner, W. J., 1

Ultraviolet Light, 3–4
Uranus, 15, 102
UVA, 3
UVB, 3–4

Venus, 13, 90
Volcanoes, 52

Water recycling, 158–9
White, Frank, 119
Whitman, Walt, 29, 39, 59, 89
Wind power, 63, 73–4